THE ARTERIAL SHIFT

Reclaiming the Surface by Moving Global Trade Underground

THE ARTERIAL SHIFT

Reclaiming the Surface by Moving Global Trade Underground

Mahendra Jagir

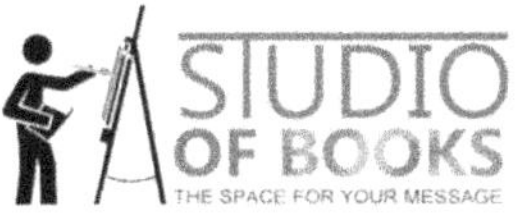

Studio of Books LLC
5900 Balcones Drive Suite 100
Austin, Texas 78731
www.studioofbooks.org
Hotline: (254) 800-1183

Ordering Information:
Special discounts are available on quantity purchases by corporations, associations, and others. For details, contact the publisher at the address above.

Printed in the United States of America.

ISBN-13: Paperback   978-1-970283-38-9
           Ebook   978-1-970283-39-6

Library of Congress Control Number: 2026905210

Dedication

This work is dedicated to my grandchildren—Dustin, Matthew, Georgia, and Jolie.

I write this so that you may grow up in a world where the horizon is defined by trees and skylines, not by walls of rusted steel containers. Where the coastlines are open, highways are quieter, and cities breathe more freely.

May you inherit a planet where the surface is reserved for beauty—and where "clutter" is only a memory of the past.

This is my gift to your future: a world reclaimed.

Preface

The Day the Surface Stood Still

There was no single headline that declared the crisis. No dramatic collapse. Instead, it accumulated quietly—ship by ship, truck by truck, container by container.

Across the world, thousands of cargo vessels began idling offshore, waiting for berth space. The engine ran continuously. Fuel burned. Carbon rose into the atmosphere. On land, port yards overflowed with stacked steel containers, spreading outward into industrial sprawl. Highways filled with diesel freight traffic. Supply chains slowed. Costs have increased.

The surface did not collapse. It simply became congested.

We have treated the Earth's surface as if it were a storage unit—an endless expanse available for stacking, parking, and warehousing the machinery of trade. But the surface is not a warehouse. It is the skin of our planet. It supports cities, ecosystems, public life, and economic vitality.

When friction builds in a mechanical system, engineers redesign the system. They do not accept inefficiency as inevitable.

The Subterranean Arterial Network (SAN) is not merely a railway proposal. It is a structural redesign. A surgical correction to a logistics model that has remained fundamentally unchanged since the steam engine. It proposes a simple shift: move the heavy muscles of trade underground and return the surface to productive, sustainable use.

This book begins with a premise grounded in practicality: global trade must continue—but it does not need to occupy the horizon.

The surface can be reclaimed.

Table of Contents

Introduction

The Silent Revolution

For more than a century, humanity has equated progress with elevation. We built taller buildings, faster aircraft, higher bridges, and digital networks that orbit the planet. Economic ambition pointed upward, and our skylines became symbols of advancement. Nations measured growth by what rose above the horizon.

Yet beneath that upward expansion, the foundation of global trade changed very little.

Containers still move from ship to dock, dock to yard, yard to truck. Highways absorb the flow. Cities absorb congestion. Emissions accumulate. Land disappears beneath warehouses, staging yards, and distribution corridors. As trade volumes expanded exponentially, the physical architecture supporting that trade remained largely surface-bound.

The result is not simply congestion. It is structural friction.

Ports that once functioned as transfer nodes now operate as storage fields. Container backlogs stretch across acres of reinforced concrete. Ships idle offshore awaiting berth availability. Diesel trucking networks extend inland in dense corridors of dependency. Prime urban and coastal land—among the most valuable real estate on earth—is consumed by logistics sprawl.

This is not a failure of trade. It is a failure of spatial design.

Modern economies depend on uninterrupted circulation. When circulation slows, costs rise. When pressure builds, inefficiencies multiply. Congestion, emissions, land loss, and supply chain volatility are not isolated problems; they are symptoms of a surface-level system operating beyond its intended capacity.

While our heads were in the clouds, our feet became entangled in infrastructure designed for a different century.

The next great infrastructure shift will not be higher. It will be deeper.

The Arterial Shift introduces the Subterranean Arterial Network (SAN), a structured reconfiguration of freight movement that relocates primary trade corridors beneath the earth's surface. Rather than expanding highways, stacking more containers, or widening port yards, SAN redirects freight into controlled, high-speed underground arteries that connect directly from maritime terminals to inland logistics hubs.

This is not a speculative exercise. The technological components required already exist in mature or emerging forms:

- High-capacity tunnel boring systems capable of sustained deep-corridor construction

- Magnetic levitation and low-friction propulsion technologies proven in commercial transport

- Vacuum-assisted transport concepts that reduce drag and energy demand

- Autonomous logistics and robotics deployed across modern ports and warehouses

- Renewable energy integration models aligned with grid-scale deployment

What has been missing is integration—an architectural framework that unifies these technologies into a coherent, scalable freight system.

SAN treats global trade as a circulatory network. When arteries are clear, circulation is efficient. When blocked, pressure builds, and strain spreads outward. Surface congestion is not merely an inconvenience; it is arterial obstruction within the economic body.

By relocating freight arteries underground, circulation stabilizes. Surface pressure declines. The economy gains flow without expanding its physical footprint.

The implications are structural.

First, environmental performance improves. Electrified subterranean corridors reduce reliance on diesel trucking and mitigate port-adjacent pollution zones. Integration with renewable energy grids aligns freight mobility with long-term climate targets. Reduced land conversion protects agricultural areas, coastal ecosystems, and urban green space.

Second, economic efficiency strengthens. Predictable transit times reduce supply chain volatility. Ports regain throughput stability. Inland hubs decentralize congestion without encouraging further sprawl. The reduction of friction translates directly into lower operating costs and improved competitiveness.

Third, urban livability increases. Reclaimed surface land becomes available for housing, transit, public space, and climate adaptation infrastructure. Freight infrastructure becomes less visually and physically intrusive. Growth no longer requires sacrificing civic space.

The dividend is measurable:

Reduced congestion.

Lower emissions.

Recovered land.

Stabilized supply chains.

Long-term infrastructure resilience.

This book presents a structured blueprint for achieving that shift. It examines engineering architecture, financial modeling, legislative pathways, public–private capital alignment, and phased global implementation strategies. It is written for leaders responsible for infrastructure investment, climate policy, urban development, and trade modernization.

The proposal does not depend on technological miracles. It depends on coordinated execution.

Every major era of economic expansion has been accompanied by an infrastructure redesign—from rail to highways to containerization.

Each shift redefined spatial logic and reshaped growth patterns for generations. The question facing policymakers today is whether the surface can continue absorbing the full burden of expanding trade.

The answer is increasingly clear.

The future of freight does not require more land. It requires better circulation.

The revolution described in these pages will not dominate skylines or produce architectural spectacle. It will move quietly beneath them. It will not compete with cities for space; it will create space for cities to thrive.

And in doing so, it will allow the surface to breathe again.

Part I

The Crisis: A Choked Surface

Global trade has expanded in scale, speed, and complexity. Surface infrastructure has not expanded at the same rate.

Ports, highways, rail corridors, and distribution zones were engineered for steady circulation—not for sustained overflow. As vessel capacity increases and cargo volumes intensify, inland evacuation remains confined to fixed road grids and finite land footprints. The geometry remains constant while the volume grows.

The result is structural pressure.

Congestion is no longer a temporary imbalance between supply and demand. It is a spatial constraint embedded within the surface framework of modern trade. Containers accumulate. Vessels idle. Warehouses spread outward. Emissions concentrate along freight corridors. Land once flexible becomes permanently allocated to logistics storage.

This section examines that condition with clarity and precision.

The chapters that follow diagnose the mechanics of surface saturation and quantify its environmental and economic cost. Together, they establish a central conclusion: the bottleneck is not operational—it is geometric.

When circulation is confined to a choked surface, friction compounds.

Understanding that constraint is the first step toward architectural reconfiguration.

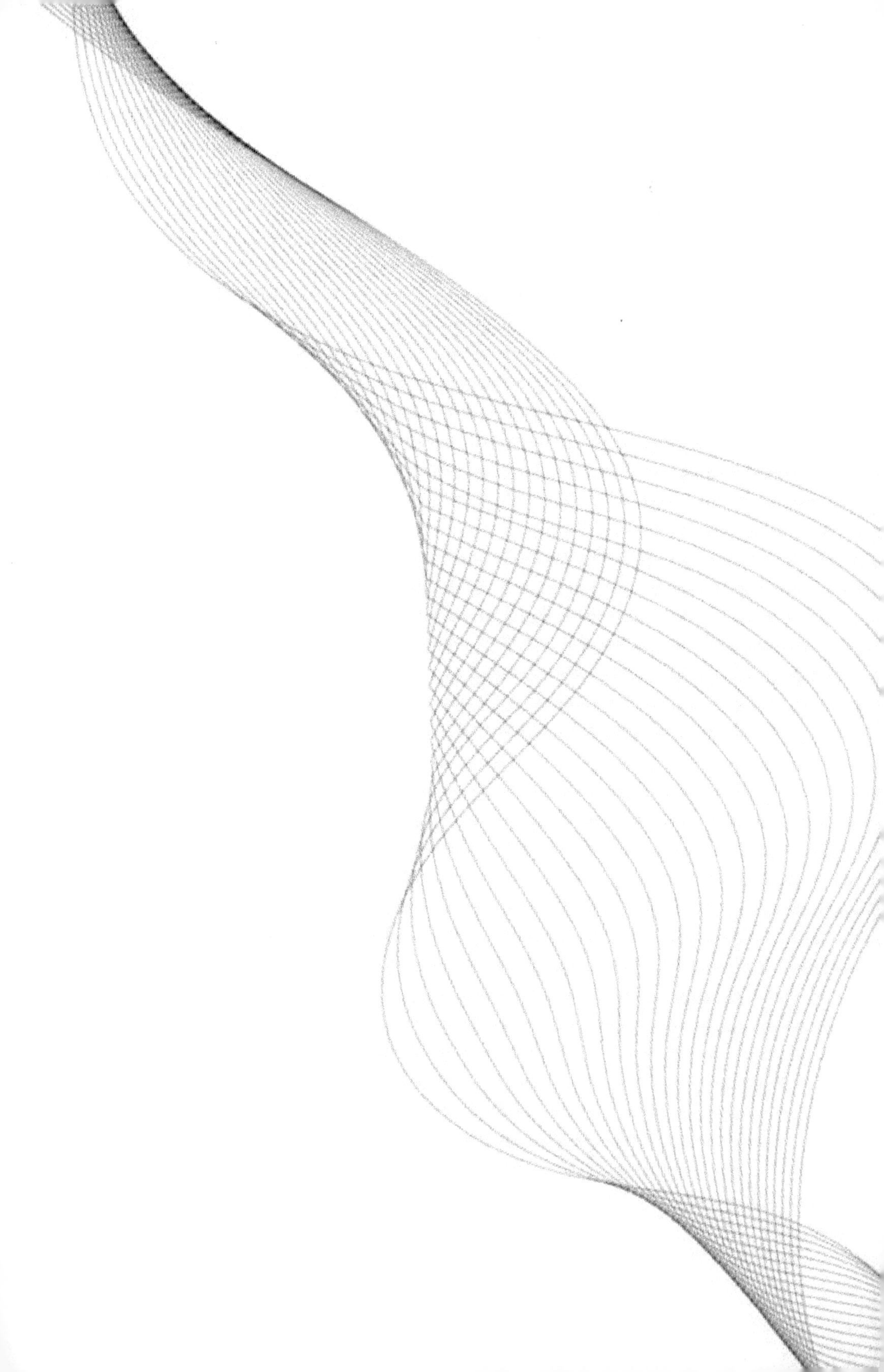

Chapter 1

The Anatomy of Port Congestion

Modern ports are engineered to move cargo, not store it. They are transitional interfaces—designed for velocity, sequencing, and precision timing. Their efficiency depends on motion. When movement slows, the system begins to accumulate pressure.Increasingly, however, ports are functioning as holding grounds for stalled circulation.

Congestion is not caused by a single operational failure. It emerges from interconnected constraints that amplify one another: container backlogs, idle vessels, yard overflow, and expanding warehouse footprints. Each condition reinforces the next. Together, they form a closed loop of surface saturation.

This chapter examines that loop—not as a temporary disruption, but as a structural condition of modern trade.

Container Backlogs

Container terminals operate on a tightly calibrated sequence: vessel arrival, crane discharge, yard staging, truck or rail departure. Every component is synchronized. The system works when each link transfers responsibility quickly to the next.

When any link slows, the entire chain compresses.

Over the past two decades, vessel sizes have increased dramatically. Ultra-large container ships deliver unprecedented volumes in single port calls. Crane productivity has improved to meet this scale, enabling faster discharge rates and higher per-hour container moves.

However, inland evacuation capacity—primarily dependent on trucking and surface rail—has not expanded at the same rate. Highways face fixed geometric limits. Rail networks compete with passenger demand and existing freight schedules. Urban land constraints restrict terminal expansion.

The result is stacking pressure.

Containers accumulate in terminal yards beyond intended dwell times. Instead of remaining for hours, they remain for days. Instead of minimal stacking, they are piled higher and denser to accommodate volume. Retrieval becomes slower as stacks grow and re-handling increases. A container buried under multiple tiers requires additional equipment moves before departure.

Each additional move consumes time, labor, fuel, and coordination. Productivity declines not because cranes slow, but because circulation stalls.

What was designed as a flow-through interface becomes a temporary warehouse operating under strain.

Backlogs are not merely operational inefficiencies. They represent blocked circulation within the economic body. When arteries narrow, pressure builds upstream. In the same way, container accumulation signals that the surface system is absorbing more volume than it can evacuate.

Idle Vessels
When berth availability is constrained by yard saturation, ships wait offshore.

A vessel that cannot discharge cannot depart. A berth occupied longer than scheduled delays the next arrival. Anchored ships form visible evidence of invisible congestion within the yard.

Idle vessels consume fuel for auxiliary systems, refrigeration units, and onboard power generation. Crews remain on extended rotations. Insurance and charter costs accumulate. Most critically, schedules destabilize.

Maritime logistics operates on sequencing discipline. Carriers design global routes based on predictable turnaround times. When one port slows, the disruption cascades across entire trade lanes.

A delayed vessel arriving in Asia or Europe affects manufacturing input timing. Retail inventory cycles adjust. Buffer stocks increase. Freight premiums fluctuate. The cost of unpredictability compounds across industries.

The economic cost of idling is therefore not confined to port authorities. It propagates through supply chains, influencing pricing structures and capital allocation decisions worldwide.

In this context, congestion is not localized. It is systemic friction embedded within global circulation.

Yard Overflow
Terminal yards were never designed for indefinite storage. They are transitional spaces—staging platforms between sea and land.

As backlogs persist, overflow expands beyond core terminal boundaries. Temporary container stacks appear in auxiliary lots. Nearby industrial parcels are leased. In some cases, open land not originally zoned for heavy logistics becomes emergency storage.

The physical footprint of congestion spreads outward, often into urban or environmentally sensitive areas.

Operational complexity increases with every expansion. Equipment must travel longer internal distances. Traffic routing inside terminals becomes less efficient. Labor coordination grows more complicated. Safety risks escalate as density increases.

What begins as a throughput delay evolves into a spatial problem.

Ports become patchworks of reactive storage zones rather than coordinated transfer hubs. The system compensates for limited inland evacuation by consuming additional surface land.

This expansion appears incremental, but its impact is cumulative.

Surface Warehouse Sprawl
Inland distribution centers were designed to complement port throughput—to receive cargo efficiently and distribute it across regional markets.

Instead, they increasingly absorb overflow.

To accommodate rising volumes and buffer uncertainty, logistics developers acquire large tracts of land near highway interchanges and metropolitan peripheries. Warehouses expand in scale and number. Single facilities now span hundreds of thousands of square meters. Truck traffic intensifies along connecting corridors.

Surface logistics land use becomes permanent.

Once agricultural or mixed-use land is converted to large-scale warehousing, it rarely returns to civic or ecological function. Road expansions follow warehouse construction. Zoning adjusts to freight priority. Residential growth patterns adapt around logistics corridors.

The geometry of the city begins to reorganize around congestion management rather than human-centered design.

This transformation carries long-term implications:
Increased diesel exposure for adjacent communities
Higher roadway maintenance costs
Fragmentation of green space
Reduced land availability for housing or public infrastructure
Warehouse sprawl is not an isolated real estate trend. It is a spatial symptom of surface saturation.

The Closed Loop of Surface Saturation
Container backlogs delay vessel discharge
Delayed discharge produces idle vessels
Idle vessels compress global schedules
Compressed schedules intensify stacking pressure
Stacking pressure triggers yard overflow
Overflow accelerates warehouse expansion
Warehouse expansion increases truck volume.
Truck volume reinforces inland bottlenecks.

The loop closes on itself.

Surface-based freight systems are reaching geometric limits. Highways cannot expand indefinitely within metropolitan regions. Rail corridors compete for finite right-of-way. Ports are constrained by water, urban adjacency, and environmental regulation.

Congestion, therefore, is not a temporary imbalance between supply and demand. It is a structural signal.

The surface has reached functional saturation.

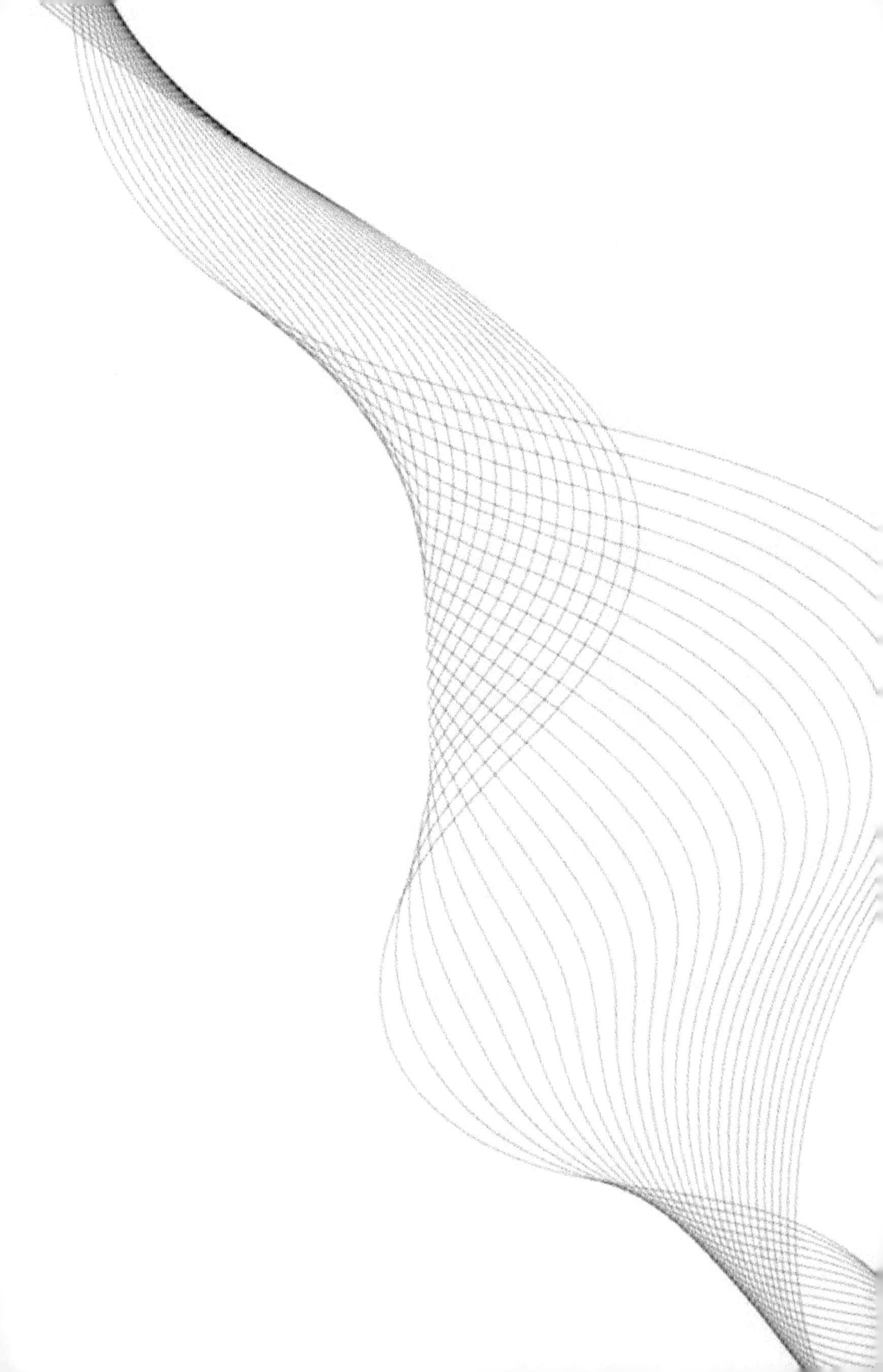

Chapter 2

The Environmental & Economic Cost

Surface-bound freight systems impose externalities that extend far beyond operational inconvenience. The costs are environmental, spatial, and financial—and they compound over time.

Congestion is often discussed in terms of delays. Yet delay is only the visible symptom. Beneath it lies a structural misalignment between trade growth and physical capacity. That misalignment produces measurable emissions, land conversion, fiscal drag, and public health exposure.

For policymakers and investors, the relevant question is not whether congestion is frustrating. It is whether the current surface-based model is economically and environmentally sustainable over decades of continued trade expansion.

This chapter evaluates that cost.

Diesel Trucking Dependency

The "last mile" of freight movement remains overwhelmingly dependent on diesel-powered trucks. Even when containers cross oceans with increasing fuel efficiency, they frequently stall once transferred to highways.

At major ports, truck queues form at terminal exits. Appointment systems and digital scheduling platforms have improved coordination, yet physical road capacity remains fixed. During peak hours, freight traffic competes with commuter vehicles along corridors never designed to function as primary global trade arteries.

The result is friction.

Diesel combustion produces particulate matter (PM2.5), nitrogen oxides (NOx), and greenhouse gas emissions. These pollutants are concentrated along freight corridors and in port-adjacent communities. Epidemiological research consistently links long-term exposure to elevated rates of respiratory illness, cardiovascular disease, and adverse developmental outcomes.

The burden is not evenly distributed. Communities located near ports and major logistics routes—often historically underserved neighborhoods—experience disproportionate exposure. Environmental inequity becomes embedded in freight geometry.

Electrification initiatives are underway. Manufacturers are producing battery-electric and hydrogen fuel cell trucks. Governments are offering incentives. Yet large-scale fleet transition requires:

- Charging infrastructure deployment

- Grid capacity upgrades

- Capital replacement cycles

- Standardization across jurisdictions

These transitions unfold over years, not months. Meanwhile, container volumes continue to increase.

Dependency on surface trucking therefore anchors trade growth to emissions growth. Even with efficiency gains, aggregate environmental impact rises as volume expands.

This dependency is not a technological failure. It is a structural one.

Maritime Emissions

When congestion prevents timely discharge, vessels wait offshore. Although modern ships are more fuel-efficient than previous generations and international maritime regulations have tightened sulfur limits, idle vessels still consume fuel for auxiliary engines, refrigeration units, and onboard systems. Engine runtime extends beyond optimal schedules. Emissions accumulate.

The environmental cost of idling is not abstract. It is measurable in additional fuel burn per vessel day and in localized air quality impact near anchorages.

Maritime decarbonization initiatives rightly focus on alternative fuels, hull efficiency, slow steaming, and propulsion innovation. These measures reduce per-unit emissions during transit.

However, efficiency at sea is undermined by inefficiency at shore.

A vessel designed for optimized voyage duration loses its environmental advantage when forced into extended anchorage. Engine hours increase without productive movement. Schedule instability can also cause vessels to increase speed on subsequent legs to recover time, partially offsetting earlier fuel savings.

In this sense, port congestion dilutes maritime decarbonization gains.

Reducing emissions therefore requires not only cleaner ships, but structural redesign of port circulation to minimize idle time.

Urban Land Consumption

Logistics infrastructure occupies some of the most strategically valuable land in metropolitan regions—coastal zones, riverfronts, highway interchanges, and peri-urban expansion areas.

As warehouse clusters multiply, residential, agricultural, and ecological uses are displaced. Large-format distribution centers require extensive footprints for building structures, truck staging areas, and surface parking. Road widenings follow. Interchange expansions follow. The freight network expands horizontally.

Land conversion becomes permanent.

Once allocated to logistics storage and truck circulation, land rarely reverts to mixed-use or public function. Surface dedication to freight is cumulative. Each additional parcel reinforces the spatial dominance of logistics over civic development.

This conversion carries opportunity cost:

- Reduced land availability for affordable housing
- Limited capacity for floodplain restoration
- Fragmented green corridors
- Constrained public transportation expansion

Urban resilience—particularly in coastal regions vulnerable to sea-level rise—competes directly with logistics expansion for limited surface space.

Continued horizontal freight growth intensifies this tension. The city adapts around congestion rather than around community need.

Land, unlike capital equipment, cannot be scaled indefinitely. It is finite.

Trade Inefficiencies
Congestion imposes financial costs across the supply chain. These costs are often absorbed incrementally, which obscures their cumulative scale.

Operational impacts include:
- Extended container dwell times
- Equipment underutilization

- Schedule unreliability
- Insurance premiums linked to delay risk
- Increased fuel consumption
- Labor overtime and re-handling expenses

Each delay may appear marginal. Yet across thousands of containers and hundreds of vessels, small inefficiencies aggregate into substantial economic drag.

Unpredictability introduces secondary costs. Manufacturers increase safety stock to hedge against delivery delays. Retailers expand buffer inventories. Working capital becomes tied up in transit uncertainty rather than productive investment.

Volatility also affects investor confidence. Infrastructure assets that cannot guarantee throughput stability introduce risk into national trade competitiveness. For export-driven economies, even modest reductions in reliability can shift routing decisions toward alternative gateways.

The friction of congestion therefore influences capital allocation.

Trade inefficiencies are not temporary anomalies. They reflect structural limits within the surface-based model.

The Structural Equation
Surface congestion generates a reinforcing cycle:

- Increased volume intensifies trucking demand.
- Trucking demand elevates emissions and roadway wear.
- Emissions generate regulatory pressure and public health costs.
- Road expansion consumes additional land.
- Land consumption limits urban flexibility and resilience.
- Delays introduce supply chain volatility.
- Volatility encourages excess storage and sprawl.

The cycle compounds over time.

The surface-based model does not fail because trade is weak. It fails because growth outpaces spatial capacity. Efficiency gains improve margins, but they do not alter geometry.

As long as freight remains constrained to two-dimensional corridors—roads, rails, and surface yards—capacity expansion will require either more land or more congestion.

Both carry escalating cost.

PART I

Strategic Summary

If congestion were merely cyclical, incremental efficiency gains would resolve it. Automation upgrades, digital appointment systems, predictive analytics, and extended gate hours would restore equilibrium. Many of these improvements are already in place—and they have delivered measurable benefits.

Yet recurring global disruptions reveal a deeper constraint.

Modern trade growth remains tethered to two-dimensional surface movement. As vessel capacity increases and cargo volumes intensify, the inland evacuation system remains confined to fixed road grids and legacy rail alignments. The geometry does not change. Only the volume does.

When volume expands within fixed geometry, friction is inevitable.

The anatomy of port congestion exposes a foundational truth: the bottleneck is spatial. Containers accumulate not because ports lack competence, but because circulation is constrained to a saturated surface.

The consequences extend beyond delay.

The environmental and economic costs of congestion are not isolated line items. They represent systemic drag within global circulation. Friction at ports propagates through supply chains, labor markets, capital allocation decisions, and urban land use.

For policymakers, this friction translates into:

- Increased public health expenditures linked to freight emissions
- Accelerated infrastructure maintenance burdens
- Gaps in climate mitigation targets
- Reduced flexibility in urban land planning

For investors and trade executives, it translates into:

- Throughput instability
- Capital inefficiency
- Insurance and financing risk premiums
- Competitive vulnerability in global routing decisions

The issue is not whether diesel engines can be marginally cleaner or warehouses marginally more automated. Those improvements are necessary, but they operate within the same constrained surface framework.

The underlying challenge is structural dependency on surface geometry.

If trade volume continues to grow—as demographic expansion, digital commerce, and industrial demand suggest it will—the environmental and economic cost of surface saturation will rise proportionally. Efficiency gains will slow the pressure. They will not eliminate it.

The anatomy of congestion reveals the blockage.
The environmental and economic analysis reveals its consequence.

Until the circulation system itself is redesigned, surface saturation will continue to compound—absorbing land, increasing emissions, and introducing unpredictability into the economic bloodstream.

This diagnosis is not an indictment of ports. It is a recognition of scale.

The next step is not incremental mitigation.
It is architectural reconfiguration.

If the surface arteries are constrained, the solution cannot remain confined to them.

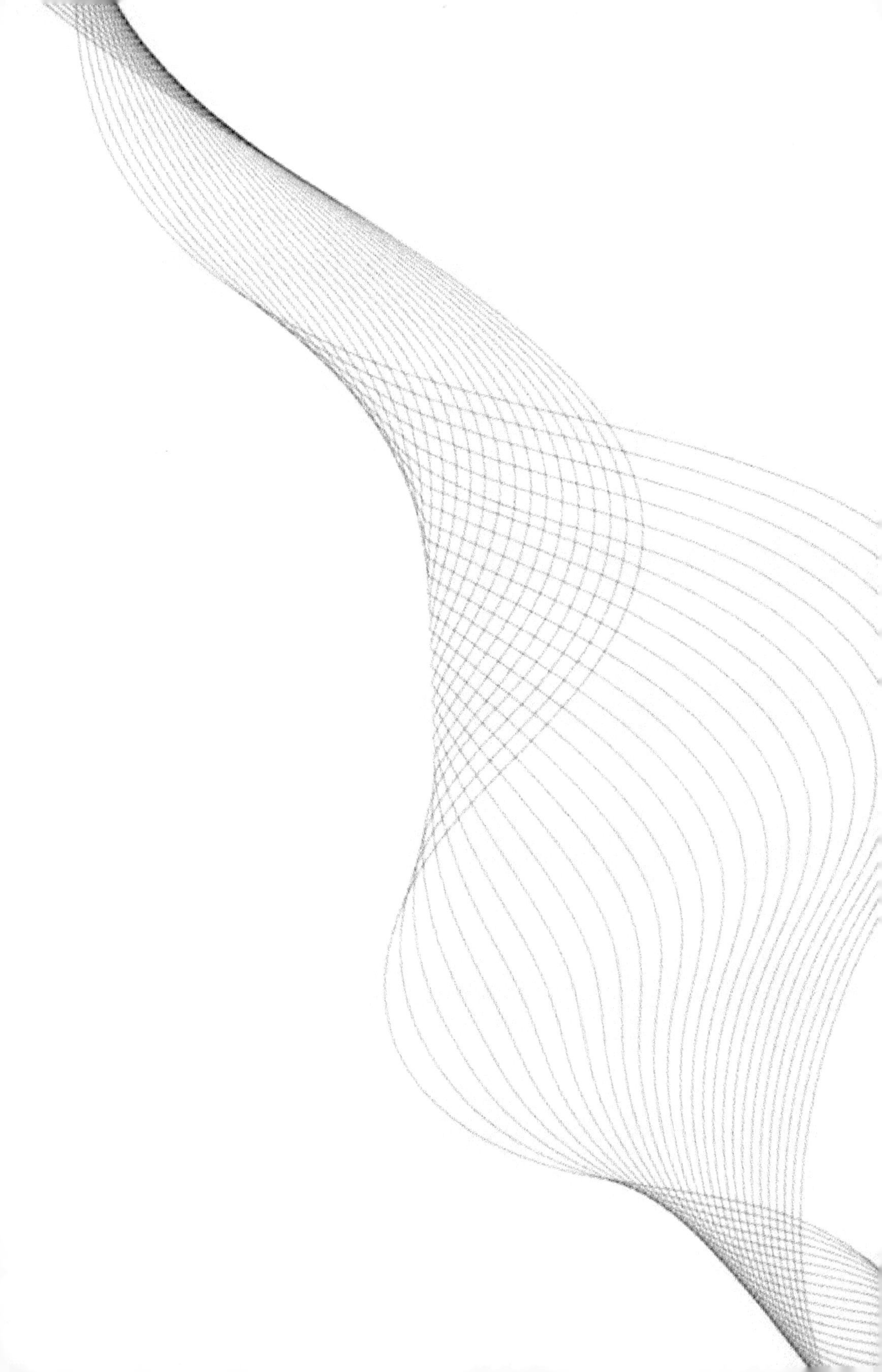

Part II

The Architectural Solution: The Subterranean Arterial Network (SAN)

If the constraint is geometric, the solution must be architectural.

Surface congestion cannot be resolved solely through incremental expansion or marginal efficiency gains. Highways cannot widen indefinitely within dense metropolitan regions. Ports cannot continue to convert scarce waterfront land into overflow storage. The geometry must change.

The Subterranean Arterial Network (SAN) introduces depth as a structural dimension of trade.

Rather than forcing growing cargo volumes to compete for finite surface corridors, SAN reallocates primary trunk movement below grade. Transfer interfaces are redesigned. Long-haul freight shifts into controlled, high-speed subterranean arteries. Redistribution concentrates within strategically positioned underground nodes.

This is not speculative engineering. It is the integration of mature technologies—vertical shaft construction, magnetic levitation propulsion, advanced tunnel boring systems, and AI-guided logistics—into a unified circulation architecture.

The chapters that follow outline this architecture in sequence:

- Redesigning the transfer point through vertical extraction.
- Establishing continuous high-speed movement beneath the surface.
- Stabilizing distribution through inland subterranean hubs.

Together, these elements form a coherent system.

Friction is removed at origin.
Velocity is sustained in depth.
Distribution is balanced at nodes.

The objective is not to relocate congestion, but to eliminate its structural cause by altering the geometry of circulation itself.

Chapter 3

Vertical Extraction Protocol

Surface congestion begins at the point of transfer.

A container ship arrives. Cranes discharge cargo efficiently. Yet once the container touches land, it enters a horizontal staging environment dependent on trucks, yard tractors, and surface sequencing. The friction is not maritime. It is terrestrial.

If congestion originates at the transfer interface, then the transfer interface must be redesigned.

The **Vertical Extraction Protocol (VEP)** establishes a direct, engineered connection between maritime terminals and deep freight corridors. Instead of discharging containers into horizontal yards for truck staging, containers move vertically into secured subterranean channels immediately upon clearance.

The objective is simple: remove the first bottleneck in the circulation chain.

Reframing the Transfer Geometry

Traditional port geometry is horizontal. Containers move from ship to crane, crane to yard, yard to truck, truck to highway. Every movement occurs within two-dimensional surface constraints.

As volume increases, the yard expands outward. Stacks grow higher. Truck lanes multiply. Surface acreage becomes the buffer for throughput volatility.

The Vertical Extraction Protocol introduces a third dimension.

By adding depth to port geometry, the staging function is relocated below grade. The surface transitions from storage platform to precision interface. Movement replaces accumulation.

This is not speculative engineering. It is geometric reallocation.

Deep-Well Docking

Deep-Well Docking integrates reinforced vertical shafts adjacent to— or beneath—existing port terminals. These shafts connect dockside crane systems directly to underground freight arteries.

Modern civil engineering demonstrates the feasibility of deep vertical access in dense urban environments. Metro systems routinely construct shafts exceeding 30–50 meters in depth. Utility corridors, stormwater tunnels, and deep foundation caissons operate safely under complex load conditions. Tunnel boring machines (TBMs) already function beneath active cities without disrupting surface activity.

Ports themselves operate heavy-lift gantry cranes capable of precise container placement within tight tolerances. The technical integration required is not invention, but alignment.

In a Deep-Well Docking configuration:

1. A container is discharged from vessel to crane.

2. Instead of being placed on a yard chassis, it is positioned onto a vertical transfer platform.

3. The platform aligns with a secured shaft aperture.

4. The container descends to a subterranean transfer chamber.

Structural reinforcement ensures load distribution. Sealed shaft systems protect against water intrusion in coastal environments. Redundant braking and stabilization systems ensure safety during descent.

This approach reduces yard dwell time substantially. Containers do not wait for trucks to become available. They enter the underground network immediately after customs clearance and digital sequencing.

Deep-Well Docking is therefore not a departure from port mechanics. It is an extension of vertical engineering principles already proven in transit, mining, and urban infrastructure projects.

Gravity-Fed Container Descent

Where geological and structural conditions allow, gravity-assisted descent systems can reduce operational energy demand.

Containers are stabilized within guided shafts equipped with:

- Automated braking systems
- Magnetic or mechanical stabilizers
- Dynamic load sensors
- Redundant safety locks

Descent occurs in controlled sequences. Regenerative braking technology—already common in elevators, electric rail systems, and industrial hoists—captures kinetic energy generated during downward movement.

Recovered energy can be redirected to:

- Shaft lighting and ventilation systems
- Automated chamber operations
- Battery storage arrays
- On-site port microgrids

This introduces partial energy circularity at the transfer stage.

Gravity, within this framework, becomes an asset rather than a constraint. It assists the movement of mass toward depth rather than requiring continuous lift-based handling at surface level.

Energy modeling would determine corridor-specific feasibility. Not all ports will implement gravity-fed systems identically. However, where applicable, this approach lowers per-container transfer energy requirements and contributes to operational efficiency.

Port-to-Depth Transfer System
At depth, containers enter sealed transfer chambers connected directly to high-speed freight corridors.

These chambers function as controlled logistical nodes. Within them, automated systems perform:

- Container identification via RFID and digital manifest integration

- Weight verification and load balancing

- Hazard classification confirmation

- Route assignment based on inland demand and corridor availability

Digital tracking systems are already embedded in global shipping networks. Containers carry standardized identifiers. Customs data platforms process documentation electronically. Satellite-linked vessel tracking provides predictive arrival data. Port management systems coordinate crane schedules with yard availability.

The Vertical Extraction Protocol integrates these digital tools into a unified routing grid.

Before a container leaves the vessel, its inland corridor pathway can be pre-assigned. Upon descent, it enters a designated lane within the subterranean artery system. Sequencing becomes continuous rather than reactive.

The transfer chamber replaces the surface yard as the primary staging environment. It operates in a climate-controlled, secure, and automated space optimized for throughput rather than storage.

Operational Advantages

Relocating the first transfer stage underground produces measurable system benefits:

- Reduced container dwell time at port
- Decreased truck queue formation
- Lower surface yard density
- Improved berth turnaround rates
- Enhanced schedule predictability

Surface congestion is reduced at its origin point.

Importantly, this redesign does not eliminate existing port labor or infrastructure. Cranes remain central. Customs processes remain intact. Surface yards continue to function as contingency buffers and specialized handling areas.

The difference lies in proportion. Storage ceases to dominate the surface footprint.

Spatial Dividend

When containers no longer require extended horizontal staging, surface acreage becomes available for reallocation.

- Ports can repurpose land toward:
- Electrified equipment deployment
- Shore power integration
- Renewable energy installations
- Storm surge mitigation systems
- Community buffer zones

The Vertical Extraction Protocol therefore produces a spatial dividend at the port interface.

This dividend compounds over time as volume scales without proportional surface expansion.

Risk Mitigation & Phased Integration

Implementation need not occur all at once.

Ports can introduce Deep-Well Docking incrementally:

- Begin with one or two high-volume berths.
- Connect shafts to a pilot corridor.
- Maintain parallel surface operations during transition.

Redundancy remains embedded in design. If underground corridors undergo maintenance, containers can temporarily revert to conventional yard staging. The system enhances resilience rather than replacing it abruptly.

Engineering risk is mitigated through phased deployment and real-time performance evaluation.

Chapter 4

High-Speed Subterranean Freight Arteries

Once below ground, freight movement transitions from intermittent trucking to continuous high-speed flow.

On the surface, cargo advances in bursts—stopping at interchanges, slowing in traffic, idling at distribution gates. Movement is constrained by weather, signal timing, road maintenance, and human variability. Even in optimized corridors, unpredictability remains embedded in the system.

Subterranean corridors redefine that geometry.

Engineered as dedicated freight arteries, underground corridors are isolated from weather variability, traffic signals, pedestrian crossings, and surface congestion. They operate within controlled environments—sealed, monitored, and optimized for constant velocity.

The objective is not novelty. It is flow.

From Intermittent to Continuous Movement
Surface trucking is inherently discontinuous. Acceleration is followed by braking. Highway speed is interrupted by bottlenecks. Even rail freight, while more efficient, shares track access and scheduling constraints.

In contrast, a subterranean freight artery functions as a closed circulation system. Containers move within pre-assigned slots, spaced digitally and propelled through uninterrupted pathways.

The difference is structural:

- Surface systems are reactive.

- Subterranean systems are sequenced.

When friction is removed, velocity becomes predictable.

Vacuum-Assisted Transport (250–300+ mph)

Air resistance is one of the primary energy costs in high-speed surface transport. At highway speeds, aerodynamic drag significantly influences fuel consumption. At higher velocities, that drag increases exponentially.

Reduced-pressure tunnel environments address this constraint directly. By operating within a controlled low-pressure atmosphere, aerodynamic resistance decreases substantially. This enables high-speed container movement with lower energy consumption per unit distance compared to equivalent surface acceleration under full atmospheric conditions.

Vacuum-assisted transport concepts have been studied extensively in both passenger and freight applications. For freight, engineering requirements are comparatively simplified:

- No human comfort constraints

- No pressurized cabin requirements

- Reduced emergency evacuation complexity

- Standardized cargo dimensions

The absence of passenger safety variables allows freight systems to focus on mechanical stability, load security, and routing efficiency.

Operating within a 250–300+ mph range enables transformative time compression. A route that might require six to ten hours by truck can be reduced to less than two hours in controlled underground transit. Longer inter-regional corridors that currently span multiple days become predictable, schedule-based movements.

Speed in this context is not spectacle. It is reliability.

Reliability stabilizes supply chains. It reduces the need for excess buffer inventory. It strengthens manufacturing sequencing. It enables precise delivery windows rather than approximate arrival estimates.

Velocity becomes a tool of economic discipline.

Magnetic Levitation Propulsion
Friction is a fundamental source of mechanical wear and energy loss in surface transport. Wheel-on-rail and tire-on-road systems require constant maintenance due to physical contact.

Magnetic levitation (maglev) propulsion eliminates this contact.

By suspending load-bearing platforms through electromagnetic fields, maglev systems remove rolling resistance. Mechanical friction declines dramatically. Maintenance intervals extend. Vibration is minimized.

Existing maglev deployments demonstrate sustained high-speed performance in controlled corridors. While most current systems are passenger-focused, the underlying propulsion technology is adaptable to freight platforms.

For freight applications:

- Load-bearing chassis can be standardized to ISO container dimensions.
- Weight distribution systems can be calibrated for heavy cargo loads.
- Autonomous control software can manage spacing and sequencing.

The absence of physical contact reduces infrastructure fatigue. Track systems experience less mechanical degradation compared to traditional rail. This extends lifecycle durability and lowers long-term maintenance expenditure.

Maglev propulsion also aligns naturally with electrified energy systems. Power is supplied directly through electrified guideways, allowing integration with renewable generation sources.

As grids incorporate greater proportions of wind, solar, and storage capacity, freight propulsion can operate within decarbonized energy frameworks.

The artery becomes electrified.

Energy Efficiency and Grid Integration
High-speed does not imply high waste.

Because subterranean corridors operate within sealed environments:

- Temperature variation is moderated.
- Crosswinds are eliminated.
- Precipitation does not affect traction.

Energy modeling indicates that continuous, high-speed electric propulsion in low-pressure environments can achieve competitive energy intensity per container-mile, particularly when compared to stop-and-go diesel trucking under congested conditions.

Regenerative braking during deceleration near inland hubs can return electricity to the grid or localized storage systems. Load-balancing algorithms can schedule departures to align with off-peak energy demand cycles.

The corridor therefore functions not only as transport infrastructure but as a predictable electricity consumer—an asset for grid stability.

Autonomous Tunnel Boring Systems

Scalability depends not only on propulsion but on construction efficiency.

Modern tunnel boring machines (TBMs) represent decades of engineering refinement. Contemporary TBMs are capable of:

- Continuous excavation
- Automated guidance and alignment
- Simultaneous segment lining installation
- Real-time geological monitoring

Advances in robotic surveying and digital twin modeling allow corridors to be mapped with precision before excavation begins. Semi-autonomous TBMs reduce labor risk and increase consistency.

Cost per mile of tunnel construction has declined in regions that apply standardized designs and streamlined permitting. Modular tunnel architecture enables phased expansion, allowing corridors to grow incrementally as demand increases.

Unlike highways, subterranean corridors do not require broad surface land acquisition. Routing can pass beneath:

- Urban districts
- Agricultural zones
- Rivers and waterways
- Environmentally sensitive areas

Surface disruption is minimized to entry and exit points.

Expansion occurs in depth rather than breadth.

This shift in geometry transforms the economics of scale. Instead of competing for scarce metropolitan land, growth leverages vertical separation.

Network Redundancy and Resilience

Subterranean freight arteries can be engineered with parallel tubes, maintenance bypass lanes, and sectional isolation gates. If one segment requires servicing, adjacent sections continue operating.

Because corridors are protected from storms, flooding, and surface accidents, service interruptions decrease. Climate volatility has limited direct impact on underground transit compared to exposed highways and rail lines.

Resilience becomes embedded in infrastructure design.

For national trade strategy, resilience is as valuable as speed. Reliable corridors protect economic continuity during extreme weather events or surface disruptions.

Environmental and Spatial Dividend

High-speed subterranean freight arteries produce secondary benefits:

- Reduced long-haul trucking demand
- Lower diesel emissions along highway corridors
- Decreased roadway maintenance due to reduced heavy truck loads
- Stabilized freight timelines

As container flow shifts underground, surface congestion declines. Ports and inland hubs require less overflow staging acreage. Urban land becomes available for alternative uses.

The surface dividend accumulates gradually but persistently.

Chapter 5

Inland Sub-Hubs – The Underground Heart

A circulatory system requires not only arteries but nodes.

High-speed corridors move containers rapidly beneath the surface, but velocity alone does not complete distribution. Flow must be received, sorted, redirected, and delivered. Without organized nodes, even the most advanced arteries would merely relocate congestion.

Inland Sub-Hubs function as the underground heart of the Subterranean Arterial Network (SAN).

Strategically positioned outside dense urban cores—yet within direct reach of metropolitan markets—these hubs serve as high-capacity redistribution centers. They connect deep freight corridors to electrified regional delivery systems, synchronizing long-distance velocity with local precision.

The objective is structural balance: speed at scale, order at nodes.

Strategic Placement and Spatial Logic
Location determines efficiency.

Sub-Hubs are positioned along major economic corridors but beyond the most land-constrained urban districts. This placement achieves three outcomes:

- Reduced surface congestion within city centers
- Lower land acquisition costs compared to coastal or central urban zones
- Direct access to existing rail, highway, and urban transit electrification infrastructure

Unlike conventional logistics parks that expand horizontally across valuable surface land, Sub-Hubs concentrate primary freight handling below grade. Surface footprints remain compact, housing administrative, maintenance, and controlled loading interfaces.

The majority of mass movement occurs underground.

This configuration allows metropolitan markets to receive high-volume freight without replicating the warehouse sprawl currently associated with surface distribution.

Robotic Container Sorting

Within each Sub-Hub, containers enter automated sorting grids engineered for precision and throughput.

Robotic gantries, conveyor platforms, and autonomous transfer vehicles coordinate movement inside climate-controlled subterranean chambers. These systems are not theoretical; they are extensions of technologies already deployed in advanced fulfillment centers and automated ports.

The difference lies in scale.

Where fulfillment centers process parcels, Sub-Hubs process full-scale containers. Standardization becomes an advantage. ISO container dimensions allow equipment calibration to uniform specifications, reducing complexity in mechanical design.

Upon arrival at depth:

1. Containers are digitally verified and scanned.
2. Routing assignments are confirmed against live demand data.
3. Units are directed to designated outbound corridors or distribution bays.

Automation reduces human exposure to heavy industrial environments while increasing throughput consistency. Robotics operate continuously without shift fatigue. Precision improves load handling. Safety incidents decline.

Containers can be:
- Transferred intact to regional distribution corridors
- Opened within controlled modules for load redistribution
- Reconfigured into smaller electric delivery units
- Directed to rail or micro-distribution centers

Importantly, these operations occur without requiring expansive surface yards. Storage density increases below grade without altering urban landscapes above.

The Sub-Hub becomes a high-efficiency redistribution chamber rather than a sprawling industrial estate.

AI Logistics Grid
At the network level, SAN operates through an integrated AI-driven logistics grid.

Modern global shipping already relies on predictive analytics for vessel scheduling and inventory forecasting. SAN extends this digital intelligence into a unified corridor management system.

The AI grid continuously evaluates:

- Real-time demand patterns
- Corridor capacity availability
- Energy pricing fluctuations
- Weather events affecting surface delivery
- Maintenance scheduling requirements
- Regional economic shifts

Dynamic allocation ensures corridor capacity is used efficiently. If one inland hub approaches peak throughput, routing can automatically adjust to alternative nodes within the network.

Load balancing reduces congestion before it forms.

This adaptive capability transforms SAN from static infrastructure into responsive mobility architecture. Rather than reacting to bottlenecks after they occur, the system anticipates and redistributes flow preemptively.

Predictability increases.

Variability declines.

For trade executives, this translates into schedule reliability.For policymakers, it reduces volatility in supply chain performance. For investors, it enhances asset utilization stability.

The AI layer does not replace physical infrastructure; it optimizes it.

Integration with Electrified Local Distribution
Sub-Hubs are designed to interface directly with electrified regional transport systems.
Outbound freight can connect to:

- Electric heavy-duty truck fleets
- Short-haul rail electrification corridors
- Urban micro-distribution centers
- Consolidated last-mile delivery networks

By concentrating high-volume trunk movement underground, surface transport shifts toward shorter-distance, electrifiable distribution.

This structural separation enables targeted electrification rather than attempting to electrify entire long-haul trucking corridors simultaneously.

Energy demand becomes localized and manageable.

Operational Resilience
Redundancy is embedded in Sub-Hub design.

Multiple access shafts, parallel sorting chambers, and sectional isolation zones ensure maintenance activities do not halt overall operations. In the event of corridor disruption, containers can be temporarily stored in controlled underground bays or rerouted to adjacent hubs.

Because primary handling occurs within sealed environments, weather variability—extreme heat, storms, heavy rainfall—has limited operational impact.

Resilience is not an afterthought. It is structural.

For national trade systems increasingly exposed to climate volatility, this durability carries strategic value.

Surface Land Reclamation Potential
As primary freight storage and transfer shift underground, surface land is released.

Port-adjacent acreage currently used for overflow stacking can transition to higher-value functions:

- Mixed-use development
- Public waterfront access
- Renewable energy installations
- Flood mitigation infrastructure
- Urban green corridors

Inland warehouse clusters can consolidate into smaller surface footprints supported by underground flow. Instead of expanding outward, logistics infrastructure stabilizes or contracts.

Urban land reclamation models demonstrate that high-density metropolitan acreage commands premium valuation when freed from heavy industrial use. Redevelopment potential increases municipal tax bases. Housing supply can expand in constrained markets. Public space can be restored.

The economic value of reclaimed land is measurable. It appears in:

- Increased property valuation
- Reduced infrastructure maintenance costs
- Enhanced resilience investment capacity
- Improved public health metrics

The surface dividend is not symbolic.

It is financial, environmental, and civic.

Long-Term Economic Implications
By relocating primary freight redistribution below grade, Sub-Hubs alter the cost structure of logistics over time.

- Land acquisition pressure declines.
- Road maintenance expenditures decrease due to reduced heavy truck traffic.
- Emission mitigation targets become more achievable.
- Throughput predictability strengthens national competitiveness.
- The underground heart stabilizes the broader system.

Instead of relying on continuous horizontal expansion to absorb trade growth, the network scales through depth and digital optimization.

PART II

Strategic Summary

The redesign begins at the point of friction.

The Vertical Extraction Protocol addresses congestion at its source: the surface transfer interface. By introducing depth into port geometry, it converts horizontal accumulation into vertical circulation. Containers descend immediately into controlled corridors rather than spreading outward across yards and highways.

This is not aesthetic transformation. It is structural recalibration.

Surface arteries are relieved because the first stage of inland movement no longer competes for scarce surface space. The port returns to its intended role—a high-efficiency interchange between maritime and inland networks. When friction is removed at the transfer point, pressure across the broader system declines.

But depth alone is insufficient without velocity.

High-speed subterranean freight arteries extend this logic beyond the port perimeter. For primary trunk movement between coastal terminals and inland logistics zones, dedicated underground corridors remove the dominant source of surface congestion. Trucks and rail continue to serve regional and last-mile functions, yet the heavy long-haul burden shifts below ground.

The engineering foundation is grounded in existing capability:

- Reduced-pressure tunnel environments

- Magnetic levitation propulsion

- Advanced tunnel boring systems

- Integrated digital routing and automation

No singular technological breakthrough is required. The innovation lies in integration—aligning mature technologies into a coherent circulation architecture.

When movement becomes continuous rather than intermittent, economic vitality strengthens. Predictability improves. Energy efficiency rises. Infrastructure wear declines.

Surface congestion is not solved by widening highways indefinitely. It is resolved by reassigning where primary flow occurs. The subterranean corridor establishes that new locus of movement.

Yet arteries alone do not sustain circulation. They require coordinated nodes.

Inland Sub-Hubs complete the architecture of SAN by synchronizing high-speed trunk transport with adaptive local distribution. These underground hubs function as the heart of the system—receiving, sorting, and redirecting freight with robotic precision and AI-guided optimization.

Complexity concentrates below grade.

The surface simplifies.

As redistribution shifts underground, urban land is freed from expansive staging yards and warehouse sprawl. The system grows without consuming the civic fabric it serves. When the heart functions efficiently, pressure equalizes across the network. Flow becomes steady. Resources move where needed without congestion shock.

The combined effect is systemic:

- Friction is removed at transfer.
- Velocity is sustained in depth.
- Distribution is stabilized at nodes.

Together, these elements form a coherent infrastructure redesign—one that aligns economic growth with spatial discipline.

The architecture does not merely move congestion from one location to another. It removes its structural cause by altering geometry itself.

The next stage of development extends this circulation model further—integrating energy systems, water reserves, and climate resilience into the same subterranean framework. Freight corridors evolve into multi-functional national infrastructure.

The arterial shift, therefore, is not about speed alone.

It is about restoring balance between trade expansion and the surface it depends upon.

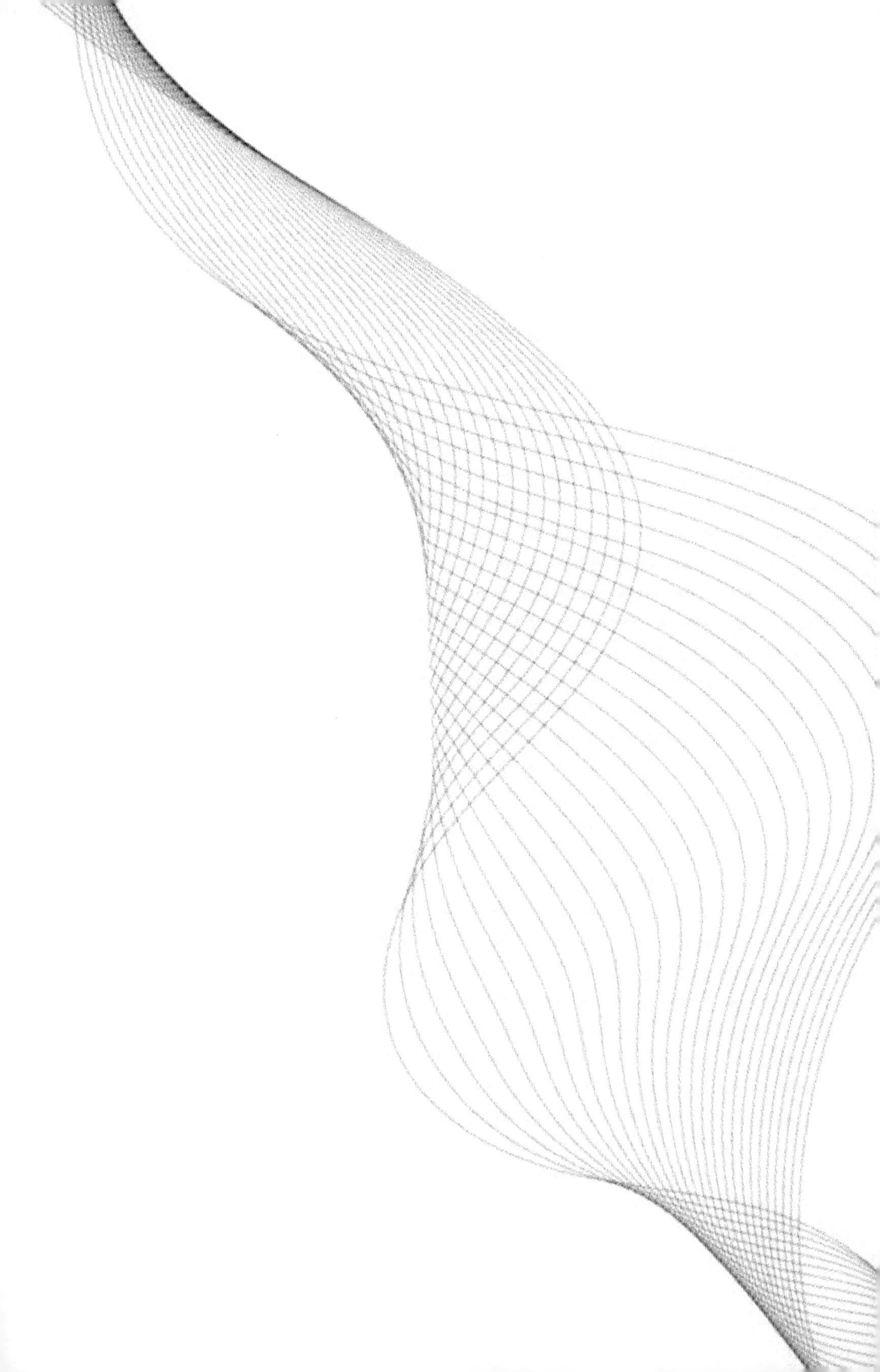

Part III

Systems Integration: SHIELD & Sustainability

Depth alone does not define resilience. Integration does.

The Subterranean Arterial Network (SAN) is not simply a transport redesign; it is a coordinated infrastructure platform. When freight arteries relocate below the surface, they create a controlled environment where energy systems, water reserves, fire suppression networks, and digital logistics can operate in alignment rather than isolation.

SHIELD represents this layered integration.

Electrified propulsion enables renewable energy pairing. Predictable corridor demand supports structured grid coordination. Subterranean alignment provides inherent protection against climate volatility. Embedded water storage and fire response systems transform transport corridors into civic resilience assets.

The objective is not to add sustainability features as peripheral enhancements. It is to embed environmental stability, energy coordination, and disaster durability directly into the operating geometry of trade.

The chapters that follow examine how renewable energy corridors, climate adaptation engineering, and resilience planning converge within a unified system.

Transport becomes more than circulation.

It becomes a stabilizing spine for energy, environment, and national continuity.

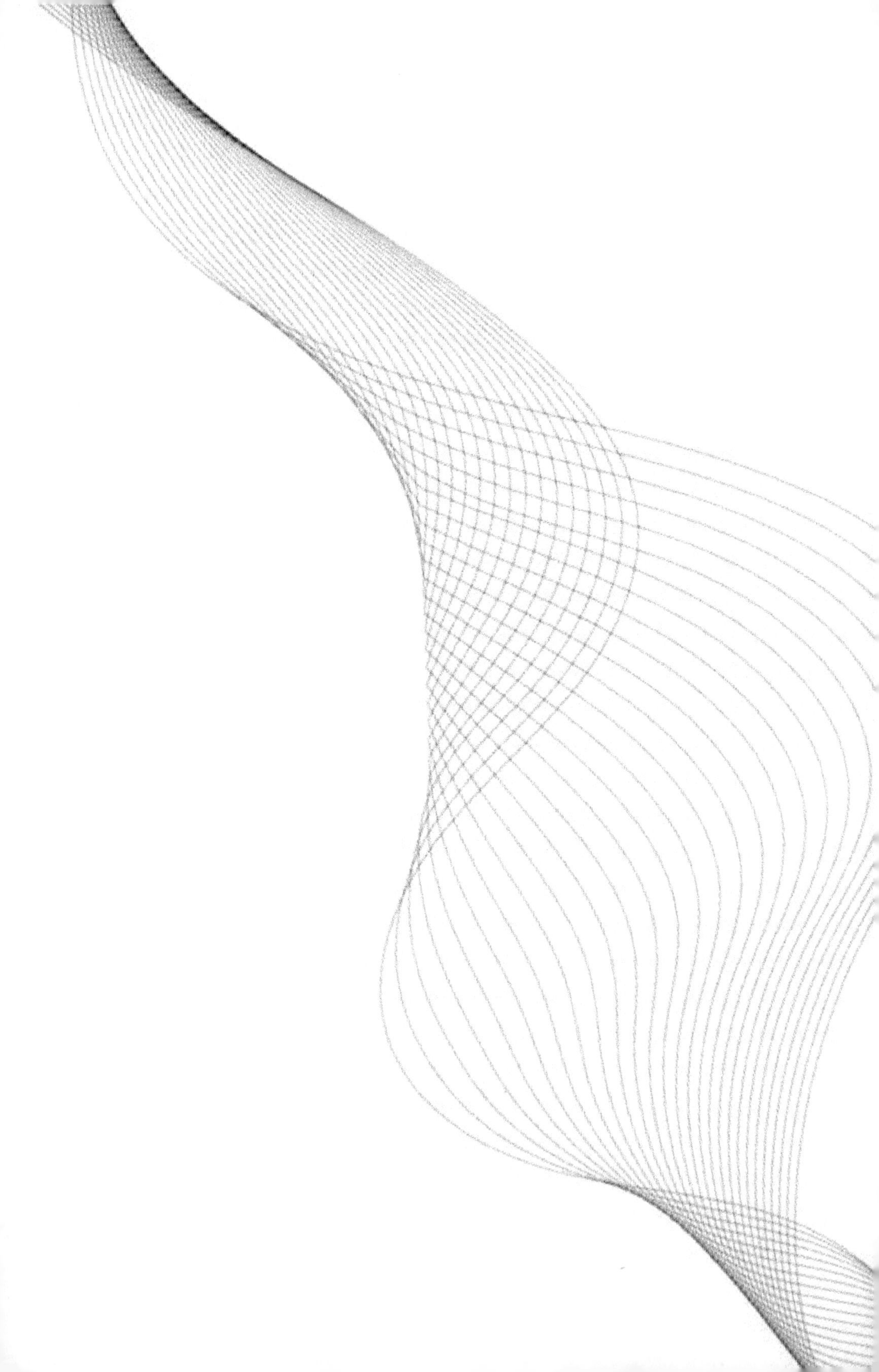

Chapter 6

Renewable Energy Corridors

Electrified freight corridors create predictable energy demand. Predictable demand enables structured renewable integration.

Traditional freight systems treat energy as an external input—diesel is purchased, consumed, and replenished with little coordination between transport demand and generation supply. The result is volatility: price exposure, emissions intensity, and limited integration with long-term grid strategy.

The Subterranean Arterial Network (SAN) repositions energy as an embedded design feature rather than a downstream dependency.

Because SAN corridors operate on fixed routes, scheduled velocity, and controlled propulsion systems, their energy profile is measurable and forecastable. This predictability creates a foundation for direct renewable integration at infrastructure scale.

Transport becomes not only a consumer of energy—but a coordinated participant in the energy ecosystem.

The Energy Logic of Electrified Corridors

Electric propulsion fundamentally alters freight economics.

Unlike diesel engines, which rely on globally traded fuel markets, electrified systems draw from regional generation portfolios. As national grids incorporate increasing shares of renewable energy, electrified freight corridors inherit that decarbonization trajectory automatically.

More importantly, SAN corridors operate with stable load patterns:

- Containers move in scheduled intervals.

- Acceleration and deceleration phases are programmable.

- Energy demand per container-mile is modelable with precision.

This structured demand allows energy planners to design renewable capacity around freight movement rather than react to it.

Infrastructure alignment replaces fuel dependency.

Wind-Powered Infrastructure

Many high-volume freight corridors naturally align with coastal regions, agricultural belts, and open-land territories where wind resources are strong and land density is lower.

Subterranean routes can be co-located with surface renewable installations—particularly wind farms—without competing for urban land. The corridor right-of-way provides a predictable linear pathway along which generation clusters and substations can be installed.

Electric propulsion systems within SAN draw directly from these regional renewable clusters. Dedicated transmission links connect wind generation fields to corridor substations. Energy storage facilities positioned along the route buffer short-term fluctuations and stabilize load requirements.

The integration is strategic:

- Wind generation peaks during certain seasonal and diurnal cycles.

- Freight corridors can adjust dispatch rates dynamically.

- Surplus energy can be absorbed through increased transport cycles or stored in battery arrays positioned at Sub-Hubs.

In effect, freight movement becomes a flexible energy consumer capable of absorbing surplus generation during peak production periods. This reduces renewable curtailment—a persistent issue in high-wind regions—and improves project economics for energy developers.

Wind generation is already a mature technology in many markets. Turbine costs have declined over the past decade. Grid interconnection standards are established. Maintenance models are well understood.

SAN does not require speculative energy breakthroughs. It leverages existing renewable infrastructure and aligns it with a predictable transport load.

Corridor-Based Energy Storage

Renewable integration requires buffering.

Energy storage systems—utility-scale batteries, pumped hydro where geography allows, or other grid-scale storage technologies—can be positioned strategically along SAN corridors and at Inland Sub-Hubs.

Storage performs three functions:

1. Absorbs excess renewable generation during peak output.
2. Supplies power during periods of low generation.
3. Provides backup capacity during grid stress events.

Because SAN energy demand is centralized along defined pathways, storage placement can be optimized for minimal transmission loss and rapid discharge capability.

Unlike decentralized electric vehicle charging dispersed unpredictably across urban networks, SAN's load is linear and concentrated. This simplifies infrastructure planning and reduces grid congestion risk.

Energy Autonomy

By integrating dedicated renewable capacity with storage and intelligent load management, SAN corridors can operate with partial energy autonomy.

Autonomy does not imply isolation from national grids. Rather, it provides resilience.

During grid stress events—heatwaves, storms, peak consumption cycles—dedicated corridor generation and storage systems allow essential freight flow to continue. This distributed energy architecture reduces vulnerability to centralized outages.

For governments concerned with energy security, this dual-use design offers strategic advantage. Freight continuity becomes less exposed to fossil fuel price shocks, geopolitical supply disruptions, or grid instability.

Energy autonomy also stabilizes long-term operating costs. Electrified propulsion decouples freight mobility from oil price volatility. Investors benefit from predictable operating expenditure models anchored in long-term renewable contracts rather than fluctuating fuel markets.

Transport transitions from a price-taker to a managed energy participant.

Grid Surplus Integration
Modern renewable systems frequently generate excess power during off-peak demand periods. Solar production peaks mid-day. Wind output often surges at night. When grid demand does not match generation, curtailment occurs—clean energy is effectively wasted.

SAN introduces a structural solution.

Because freight movement is digitally scheduled, transport velocity can be aligned with energy availability. An AI-driven logistics grid—integrated with energy market signals—coordinates routing schedules with generation patterns.

When renewable production peaks:
- Corridor throughput can increase.

- Additional container dispatch cycles can be scheduled.

- Energy storage systems can be replenished.

When production dips:

- Operations can stabilize at baseline throughput.

- Stored energy can maintain continuity.

- Non-urgent freight can be sequenced for later dispatch.

This bidirectional coordination strengthens national grids rather than burdening them.

Transport becomes a balancing asset within the broader energy ecosystem—a controllable demand resource capable of smoothing fluctuations rather than amplifying them.

Environmental and Economic Alignment

The integration of renewable energy corridors produces compound benefits:

- Reduced lifecycle emissions for freight movement

- Lower exposure to fossil fuel price volatility

- Improved return on investment for renewable developers

- Increased resilience against grid instability

- Enhanced credibility of national decarbonization strategies

Environmental and economic incentives align.
Rather than treating decarbonization as an added compliance cost, SAN embeds renewable integration within its structural logic. The corridor becomes both a transport artery and an energy conduit.

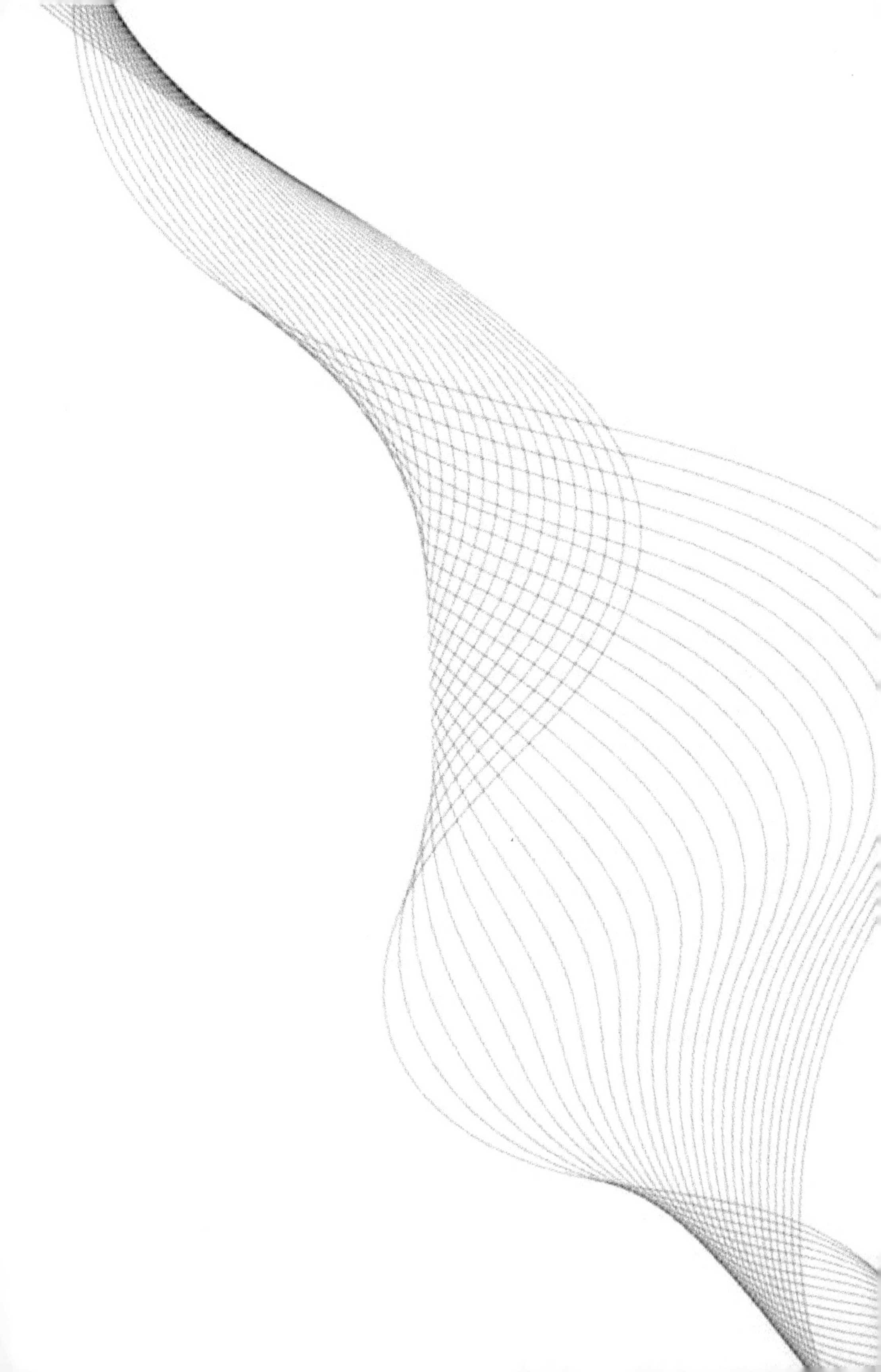

Chapter 7

Climate & Fire Protection Integration

Climate risk is no longer a secondary planning variable. It is a primary determinant of infrastructure viability.

Flooding disrupts ports. Wildfires close highways. Heat stress degrades pavement and rail systems. Storm surge compromises coastal logistics. Supply chains—once optimized solely for speed and cost—are now evaluated against durability under stress.

Surface-bound freight networks are exposed by design. They share right-of-way with weather systems, wildfire corridors, floodplains, and rising temperatures.

Subterranean corridors introduce a controlled operating environment with inherent protection advantages.

The Subterranean Arterial Network (SAN) incorporates climate resilience not as an add-on, but as an embedded design principle. Its depth is not only a congestion solution—it is a stability strategy.

Infrastructure in a Climate-Constrained Era
Modern infrastructure must perform across decades of environmental volatility. Planning horizons now extend into scenarios shaped by:

- Sea-level rise

- Increased precipitation intensity

- Extended drought cycles
- Expanding wildfire zones
- Prolonged heat waves

Traditional surface logistics systems face cumulative exposure. Every extreme event produces cascading delays, insurance losses, and economic disruption.

Subterranean systems fundamentally alter this exposure profile.

Operating below grade reduces vulnerability to wind damage, surface flooding, extreme heat, and ember-driven wildfire spread. With proper drainage engineering, reinforced linings, and pressure-controlled environments, underground corridors maintain stable operating conditions independent of surface volatility.

In this context, SAN is not merely a mobility upgrade. It is adaptive infrastructure.

Water Reserve Systems

Water scarcity and water volatility are twin challenges. Some regions face chronic drought; others experience intense flooding events that overwhelm surface retention systems.

Subterranean freight corridors offer a structural opportunity: integrated water reserve chambers positioned along strategic segments of the network.

These chambers function as:

- Emergency municipal reserves
- Agricultural buffering systems
- Industrial supply stabilizers
- Wildfire suppression resources

Deep underground storage reduces evaporation losses compared to surface reservoirs. It also minimizes land acquisition requirements in high-value urban regions.

Water can be stored in reinforced vaults adjacent to freight corridors, with dedicated pumping systems linking reserves to municipal distribution networks. In flood-prone regions, excess stormwater can be redirected into controlled underground storage, reducing pressure on surface drainage infrastructure.

By embedding water infrastructure within freight corridors, governments consolidate capital investment into a single multi-use asset. Rather than constructing standalone reservoirs, fire pipelines, and freight tunnels independently, SAN enables co-location.

Hydrological integration transforms transport tunnels into strategic reserves.

The corridor becomes a linear infrastructure spine serving both economic and civic stability.

Wildfire Suppression Corridors

Wildfires increasingly intersect with transportation routes. Highway closures during major fire events isolate regions, disrupt trade, and complicate evacuation logistics.

Subterranean corridors offer structural advantages during wildfire crises:

1. **Protection from direct flame exposure**
2. **Shielding from wind-driven embers**
3. **Stable operating temperatures below combustion zones**

Parallel to freight arteries, SAN corridors can house high-capacity water pipelines engineered for rapid surface deployment. Pressurized systems, connected to underground storage chambers, allow emergency services to access large water volumes without relying solely on distant reservoirs.

Access shafts positioned at strategic intervals serve dual functions:
Maintenance and ventilation for freight operations
Secure staging points for emergency firefighting deployment
In high-risk regions, surface firebreak zones can be aligned with underground pipeline routes. When fire activity escalates, water delivery becomes immediate and distributed rather than centralized and delayed.

Because the freight corridor itself remains insulated from surface combustion, core economic flow continues even during severe wildfire events.

The corridor becomes both an economic artery and a resilience spine.

Climate Resilience Modeling
Forward-looking infrastructure investment demands scenario modeling across decades.

Sea-level rise projections influence port viability. Heat stress impacts rail expansion joints and asphalt deformation. Precipitation variability challenges drainage systems. Population migration alters freight demand patterns.

Subterranean freight systems provide inherent climate advantages:

- Reduced exposure to extreme heat, improving mechanical efficiency

- Protection from storm surge and high winds

- Insulation from surface flooding when engineered with pressure-resistant linings and pump systems

- Stable ambient temperatures that extend equipment lifespan

Operating temperatures underground fluctuate minimally compared to surface environments. This stability reduces material fatigue, lowers maintenance cycles, and improves long-term asset performance.

Resilience modeling should therefore evaluate SAN not only as a transport upgrade, but as climate adaptation infrastructure.

When primary freight arteries are relocated below volatile surface layers, national logistics systems gain structural durability.

Depth becomes a long-term hedge against environmental uncertainty.

Environmental Emission Reduction
Climate resilience and emission reduction operate in parallel.

Electrified propulsion eliminates diesel combustion within primary freight corridors. Renewable energy integration further lowers lifecycle emissions. Reduced truck traffic decreases highway congestion and idling.

The structural redesign produces compound environmental effects:
- Surface truck volumes decline
- Idle vessel time decreases due to accelerated port clearance
- Warehouse sprawl slows as underground hubs consolidate storage footprints
- Land previously used for freight staging transitions to green or renewable uses

Importantly, emission mitigation emerges from system design—not behavioral mandates.

Trucking firms do not need to change driving habits. Consumers do not need to alter purchasing patterns. Ports do not need to reduce throughput.

The infrastructure itself reduces friction and fuel intensity.

For climate policy strategists, SAN aligns trade modernization with national decarbonization targets. Freight transitions from a high-emission sector into a managed, electrified backbone compatible with long-term carbon reduction frameworks.

The redesign converts logistics from a liability within climate accounting into a contributor toward mitigation goals.

Land Reclamation Strategy
Surface congestion consumes land.

Ports maintain overflow container yards. Inland regions develop expansive warehouse clusters. Highway expansions widen concrete footprints year after year.

As freight transfer and storage shift underground, surface acreage is released.

This land can be reallocated to:

- Urban housing development

- Public green space

- Renewable energy installations

- Floodplain restoration

- Transit-oriented infrastructure

Urban land economics consistently demonstrate that metropolitan acreage commands premium valuation when freed from heavy industrial use. Mixed-use redevelopment, tax base expansion, and increased residential density generate long-term fiscal returns.

The cost of relocating freight infrastructure underground must therefore be evaluated against the recoverable economic value of reclaimed surface land.

The surface dividend is measurable in both fiscal and social terms.

Land once consumed by congestion becomes civic capital.

Disaster & Fire Resilience Network

Beyond wildfire suppression, SAN corridors serve as hardened pathways for emergency logistics.

During hurricanes, floods, or extreme weather events, highways frequently become impassable. Rail lines may buckle under heat stress. Surface gridlock delays emergency response.

Underground corridors—engineered for controlled drainage, reinforced structural integrity, and independent power support— maintain continuity when surface systems fail.

Medical supplies, food shipments, fuel components, and essential materials can move uninterrupted between regions.

This redundancy increases national preparedness without constructing parallel emergency-only infrastructure. The freight corridor doubles as a civil resilience backbone.

Governments routinely invest in defense systems to protect territorial integrity. SAN expands that logic to economic integrity.

Freight infrastructure becomes part of civil defense planning.

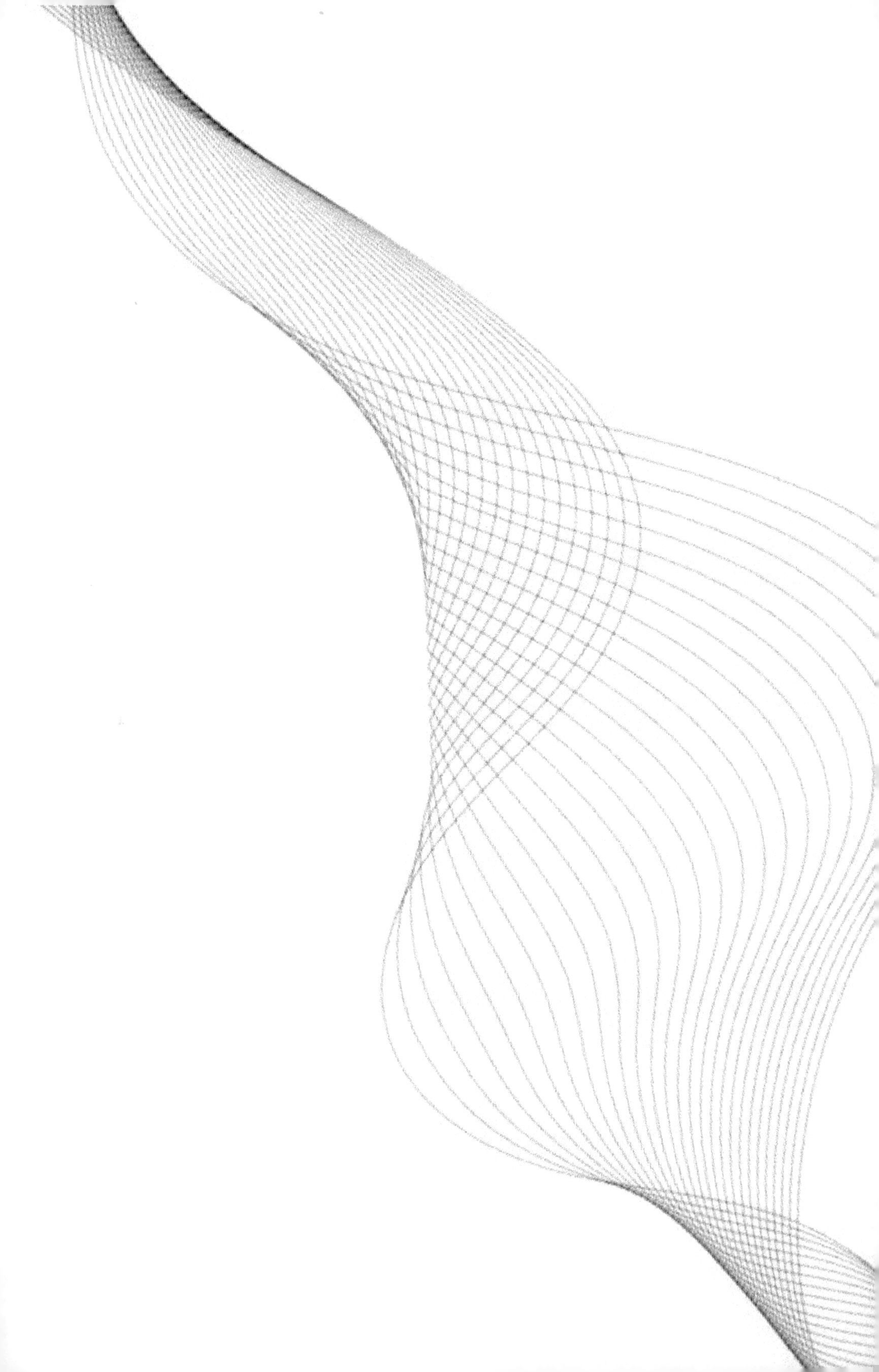

PART III

Strategic Summary

Electrified freight corridors deliver more than emission reduction. They establish predictable energy demand profiles that enable disciplined renewable integration at infrastructure scale. Wind-powered generation, corridor-based storage, partial energy autonomy, and AI-coordinated grid alignment transform freight from an energy liability into a stabilizing national asset.

The innovation is not a new fuel source. It is architectural alignment—integrating transport demand with renewable generation through intelligent infrastructure design.

As trade volumes expand, energy consumption will expand with them. The strategic question is whether that growth remains tethered to fossil volatility or transitions into coordinated renewable systems embedded within core infrastructure. SAN positions freight mobility firmly within the latter framework.

This integration represents a broader architectural shift. Infrastructure no longer performs a single function. It becomes layered and mutually reinforcing:

- Transport artery
- Energy conduit
- Water reserve
- Fire suppression network
- Disaster resilience spine

Each layer strengthens the others. When arteries operate below volatile climate zones, operational friction declines. When water reserves are embedded within transport corridors, capital efficiency improves. When renewable energy powers propulsion, emissions fall and cost predictability rises.

SAN does not attempt to outpace climate change through reactive adaptation. It redesigns the structural geometry of trade to function within a climate-constrained world. The engineering components—advanced tunneling systems, electrified propulsion, utility-scale storage, renewable generation, digital logistics platforms—already exist. The advancement lies in integration, coordination, and scale.

By relocating the backbone of freight beneath the surface, SAN protects economic circulation while restoring environmental balance above it. The surface dividend becomes measurable in land value, emission reduction, and civic resilience.

The next chapter transitions from engineering architecture to implementation strategy—examining governance frameworks, financing models, and public-private coordination mechanisms required to translate this integrated infrastructure blueprint into operational reality.

Part IV

Financial & Policy Blueprint

Engineering redesign establishes possibility. Financial architecture determines implementation.

The Subterranean Arterial Network (SAN) is not a conceptual exercise in infrastructure theory. It is a capital-intensive system that must justify investment within real-world fiscal constraints, regulatory environments, and institutional risk frameworks.

Surface congestion already extracts value through delay, emissions, land consumption, and volatility. The question is not whether capital will be deployed. It is whether it will continue funding incremental mitigation—or shift toward structural recalibration.

This section addresses that choice directly.

First, it quantifies the cost of inaction—the recurring Clutter Tax embedded in today's trade geometry. Then it outlines a disciplined revenue model anchored in subscription-based throughput, green capital alignment, and phased deployment. Finally, it examines the legislative and regulatory foundations required to convert subterranean corridors into authorized, financeable public utilities.

If earlier sections defined the engineering logic of depth, this section defines the economic and governance logic of execution.

Infrastructure becomes viable when geometry, capital, and law operate in alignment.

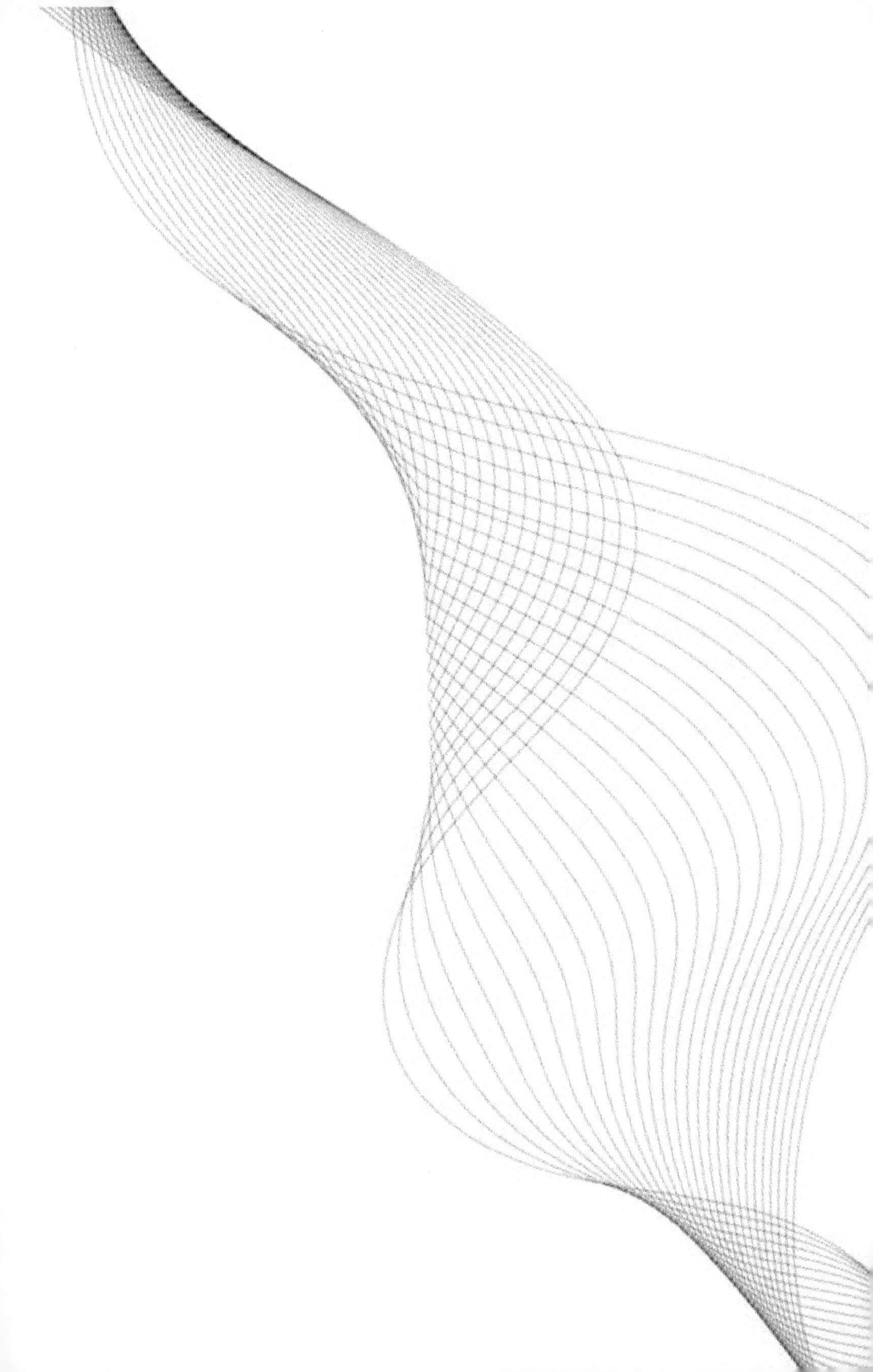

Chapter 8

The Cost of Inaction – The Clutter Tax

Surface congestion is often treated as an operational nuisance—an unfortunate but manageable byproduct of economic growth.

It is not.

It is a recurring economic penalty embedded within the geometry of modern trade.

This recurring penalty can be described as the Clutter Tax—the accumulated cost imposed on economies by surface saturation.

Unlike a formal tax, it is not legislated. It does not appear as a single line item in national budgets. Yet it extracts value continuously from ports, manufacturers, retailers, consumers, and governments.

It is paid in delay, inefficiency, emissions, land consumption, and lost opportunity.

And it compounds annually.

The Anatomy of the Clutter Tax
The Clutter Tax emerges when expanding trade volume is forced to circulate within fixed two-dimensional infrastructure.

Road grids do not scale proportionally with vessel capacity. Urban land does not expand alongside container throughput. Rail corridors share right-of-way with passenger demand. Warehousing clusters sprawl outward into increasingly distant acreage.

When volume increases within fixed geometry, friction is inevitable.

That friction translates into measurable inefficiencies:

- Extended port dwell times
- Vessel queuing delays offshore
- Increased fuel consumption during idling
- Truck idle hours in terminal queues
- Labor inefficiencies tied to unpredictable scheduling

Expanded inventory buffers to hedge against volatility
Individually, each inefficiency appears marginal. Aggregated across global trade volumes, they compound into significant annual losses.

The Clutter Tax is therefore not episodic. It is structural.

Annual Global Inefficiencies

Global trade operates on narrow margins and tight schedules. Even small disruptions ripple across supply chains.

When containers remain at ports longer than scheduled, downstream systems adjust defensively:

Manufacturers increase inventory holdings to avoid stockouts. Retailers incorporate delay premiums into pricing structures. Shipping lines incur demurrage and detention fees. Trucking firms absorb idle labor costs. Port authorities invest repeatedly in incremental surface expansions—additional staging yards, expanded gate complexes, extended operating hours—that deliver diminishing returns over time.

Each adjustment is rational in isolation. Collectively, they signal systemic inefficiency.

The financial impact includes:

- Working capital tied up in excess inventory
- Increased insurance premiums linked to volatility
- Higher transportation fuel expenditure
- Overtime labor and scheduling inefficiencies
- Recurrent capital expenditure on surface widening projects

These are not speculative costs. They are observable across major trade corridors during peak congestion cycles.

When aggregated across national economies, the cumulative effect becomes structural drag on GDP growth.

Growth does not stall because demand is absent. It slows because circulation is constrained.

Diminishing Returns of Surface Expansion
The traditional response to congestion is expansion:

- Widen highways
- Add truck lanes
- Increase terminal acreage
- Extend operating hours
- Automate surface yards

These measures produce short-term relief. Over time, however, they encounter spatial limits and induce additional volume.

Highway widening consumes land and often generates new traffic demand. Port expansion into adjacent urban zones increases land acquisition costs and community resistance. Warehouse sprawl lengthens last-mile distances and increases emissions.

Incremental surface expansion becomes progressively more expensive while delivering progressively smaller improvements in throughput stability.

The geometry remains unchanged.

The Clutter Tax persists.

Environmental Penalties
Surface-bound freight systems generate environmental liabilities that are increasingly priced into economic frameworks.

Congestion produces:

- Air quality degradation in port-adjacent communities

- Elevated particulate exposure for vulnerable populations

- Rising carbon accounting exposure for corporations

- Regulatory compliance costs linked to emissions standards

- Accelerated infrastructure wear requiring continual resurfacing and repair

As carbon pricing mechanisms expand and environmental disclosure requirements tighten, emissions are no longer abstract externalities. They become explicit financial variables.

Companies face pressure from investors to reduce Scope 3 emissions. Governments integrate decarbonization targets into trade policy. Insurance markets adjust risk premiums in climate-vulnerable regions.

Inaction compounds two penalties simultaneously:

1. Economic inefficiency from congestion
2. Regulatory exposure from emissions and environmental degradation

The longer the surface model persists without structural redesign, the higher the cumulative liability.

The Clutter Tax grows both economically and environmentally.

Land Consumption and Opportunity Cost

Surface congestion also consumes one of the most finite resources in metropolitan regions: land.

Container overflow yards, truck staging areas, and expansive warehouse clusters occupy acreage that could otherwise serve higher-value uses:

- Housing development
- Public infrastructure
- Green space
- Renewable installations
- Flood mitigation systems

Land tied up in low-efficiency logistics activity carries opportunity cost.

Urban land markets consistently demonstrate that reclaimed acreage can generate substantial tax revenue and civic value when redeveloped strategically. When freight systems require ever-expanding surface footprints, cities sacrifice long-term economic flexibility.

The Clutter Tax therefore includes not only operational inefficiency, but forgone civic capital.

Lost GDP Potential

Infrastructure does not merely support growth—it determines its ceiling.

When freight throughput is constrained by congestion, trade expansion is capped regardless of demand strength.

Delays discourage manufacturing investment. Companies hesitate to site production facilities in regions where logistics unpredictability threatens supply continuity. Exporters face reputational risk when delivery timelines become unreliable.
National competitiveness weakens gradually rather than abruptly.

Lost GDP potential rarely appears in annual budget sheets. It manifests over time in slower productivity growth, reduced foreign direct investment, and diminished trade resilience.

The Clutter Tax is therefore not a visible crisis. It is a quiet limiter.

It constrains scale without announcing itself.

Volatility as Structural Risk

Global trade volatility—exposed during major disruption cycles—reveals the fragility of surface-dependent systems.

When a shock occurs—whether pandemic, labor dispute, extreme weather event, or demand spike—surface congestion amplifies the disruption.

Buffers prove insufficient. Queues lengthen. Inventory shortages cascade across sectors.

A structurally saturated system has limited elasticity.

The Clutter Tax includes this reduced shock-absorption capacity.

Economies with constrained circulation recover more slowly from disruption. The cost of delay multiplies during crisis periods.

Structural fragility becomes macroeconomic risk.

The Strategic Financial Question

The financial question facing policymakers and investors is not whether congestion is costly. It is whether continued incremental surface expansion yields greater long-term return than structural redesign.

Surface widening projects require recurring capital infusion. Maintenance costs rise. Environmental compliance expenses escalate. Urban resistance intensifies.

Each new expansion attempts to relieve friction within unchanged geometry.

Structural redesign—introducing vertical circulation and subterranean trunk movement—changes the geometry itself.

The comparison is therefore not between cost and no cost.

It is between:

- Paying the Clutter Tax indefinitely
- Investing in architectural recalibration

One is recurring and compounding. The other is capital-intensive but structurally corrective.

The Compounding Nature of Inaction
The Clutter Tax is cumulative.

Each year of inaction adds:

- Additional land consumed
- Additional emissions released
- Additional volatility priced into supply chains
- Additional working capital immobilized in inventory buffers
- Additional infrastructure wear requiring repair

Compounding works in reverse when friction is removed. When circulation improves, capital efficiency increases. When predictability rises, inventory buffers shrink. When emissions decline, regulatory exposure falls.

The cost of inaction is therefore not static.

It escalates as trade volume grows within fixed geometry.

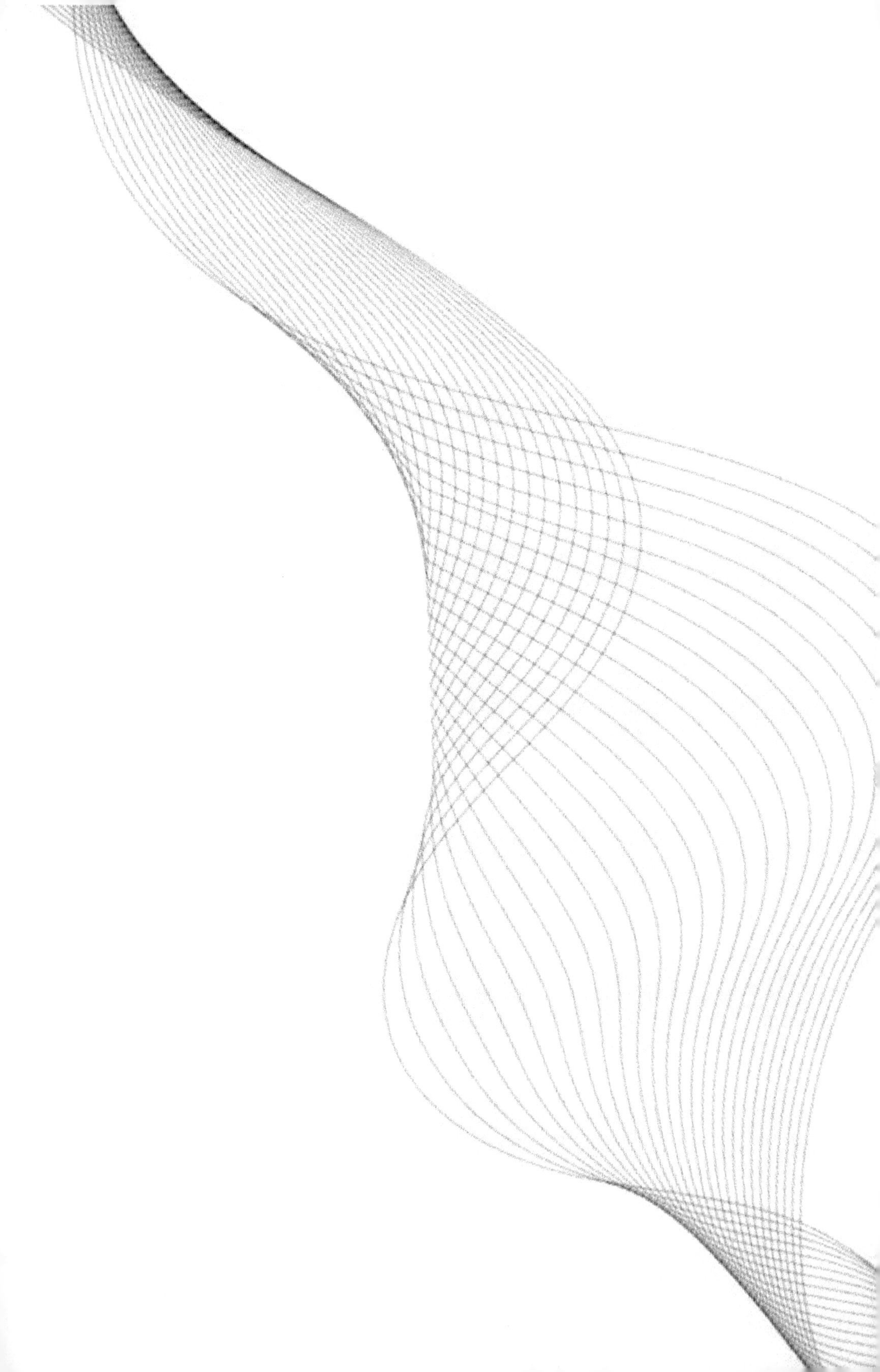

Chapter 9

Revenue & Break-Even Framework

Vision requires capital. Capital requires clarity.

Large-scale infrastructure succeeds not on engineering ambition alone, but on disciplined financial architecture. For the Subterranean Arterial Network (SAN) to move from conceptual framework to deployable system, it must demonstrate revenue logic capable of attracting both public and private investment.

The objective is not speculative return. It is durable, infrastructure-grade yield supported by long-term demand fundamentals.

SAN's financial architecture combines:

- Usage-based revenue

- Time-certainty premiums

- Green capital instruments

- Public–private alignment

- Phased capital deployment

- Quantifiable avoided public costs

The model reflects established infrastructure finance principles—applied to a reconfigured freight geometry.

Infrastructure as a Yield Asset

Global infrastructure funds, pension pools, and sovereign wealth institutions allocate capital toward assets that provide:

- Predictable cash flow

- Long-term contractual stability

- Inflation-linked revenue potential

- Low correlation with equity volatility

Traditional toll roads, regulated utilities, ports, and rail networks meet these criteria because they monetize essential services.

Freight circulation is essential. What has been missing is a mechanism to monetize reliability and congestion relief at structural scale.

SAN introduces that mechanism.

The Subscription-to-Speed Model

At the core of SAN's revenue structure is a predictable pricing framework: prioritized throughput access.

Rather than charging solely by distance moved, SAN monetizes time-certainty and congestion avoidance.

Shipping lines, logistics operators, and large-scale manufacturers subscribe to guaranteed transit speeds within defined corridors. Contracts specify:

- Maximum transit time thresholds

- Guaranteed departure intervals

- Reliability metrics backed by performance standards

This model reflects an existing market reality. In global trade, reliability commands premium value. Manufacturers routinely pay more for expedited shipping. Retailers pay to avoid stockouts. Logistics firms charge higher rates for guaranteed delivery windows.

SAN formalizes this premium within infrastructure itself.

The Subscription-to-Speed model creates:

- Recurring contractual revenue
- Multi-year service agreements
- Indexed pricing tied to volume tiers
- Reduced exposure to spot-market volatility

Revenue becomes stable and forecastable—an attractive profile for long-duration infrastructure investors.

Instead of toll variability linked to daily traffic, SAN secures contracted throughput commitments.

Reliability becomes monetized capacity.

Tiered Access and Capacity Allocation
Corridor capacity can be structured into defined tiers:

- Premium guaranteed-speed slots
- Standard scheduled throughput
- Dynamic capacity allocated during surplus periods

This tiered model optimizes asset utilization while protecting high-value contracts.

During periods of elevated renewable generation or lower corridor demand, additional capacity can be released dynamically—maximizing revenue capture without compromising reliability guarantees.

Digital logistics coordination ensures allocation transparency and operational discipline.

The corridor operates not as a static tunnel, but as a managed mobility platform.

Green Bonds and Sustainable Finance

SAN's integration of electrified propulsion, renewable energy alignment, emission reduction, and land reclamation qualifies it for green bond financing frameworks.

Global sustainable finance markets have expanded rapidly as institutional investors seek assets aligned with climate mandates and environmental, social, and governance (ESG) criteria.

SAN meets multiple sustainable infrastructure criteria:

- Direct emission reduction

- Renewable energy integration

- Urban land reclamation

- Climate resilience infrastructure

- Reduced particulate exposure in port-adjacent communities

Green bond issuance tied to specific corridor segments enables phased capital deployment. Each segment can be structured as:

- Asset-backed revenue bonds

- Climate-linked performance bonds

- Long-duration infrastructure notes

Investors gain access to stable returns aligned with environmental mandates. Governments demonstrate measurable decarbonization progress without relying solely on subsidies.

The alignment between infrastructure revenue and climate capital reduces cost of capital over time.

Public–Private Partnerships (PPPs)

Given the scale of investment required, blended capital structures are essential.

Public–private partnerships distribute risk across stakeholders while preserving performance accountability.

A structured PPP framework for SAN would allocate roles as follows:

Governments:
- Provide regulatory clarity
- Secure right-of-way access
- Establish permitting frameworks
- Offer limited revenue guarantees during early deployment

Private Consortia:
- Design and construct corridor segments
- Manage operations and maintenance
- Implement automation and digital logistics systems

Institutional Investors:
- Supply long-term capital
- Hold revenue-backed securities
- Participate in asset refinancing once operational stability is demonstrated

This model mirrors established precedent in:
- Energy transmission grids
- High-speed rail systems
- Toll highway networks
- Major port terminals

SAN extends this framework into subterranean freight infrastructure.

By anchoring revenue in subscription contracts rather than speculative demand, PPP structures reduce uncertainty and improve creditworthiness.

Phased Deployment Strategy
Infrastructure risk declines when deployment is sequenced.

SAN corridors should begin with high-density port-to-inland routes where congestion costs—and therefore reliability premiums—are highest.

Phase One characteristics:
- Major maritime port to primary inland logistics hub
- High existing truck congestion
- Established manufacturing and distribution demand
- Strong renewable energy alignment

Revenue generated from early corridors supports:
- Debt servicing
- Proof-of-concept validation
- Refinancing at lower capital cost
- Expansion into secondary corridors

Phased deployment reduces upfront capital exposure and builds investor confidence incrementally.

10-Year Break-Even Framework

Infrastructure investors require clarity on break-even timelines.

A structured 10-year break-even model for initial corridors assumes:

1. **Phased capital deployment** – Construction sequenced to limit early interest burden.

2. **Escalating subscription adoption** – Early anchor tenants commit to long-term contracts, followed by gradual onboarding of additional users.

3. **Operational efficiency gains** – Automation reduces labor intensity and improves throughput stability.

4. **Renewable energy integration** – Long-term power purchase agreements stabilize energy costs.

5. **Refinancing post-stabilization** – Once operational metrics are proven, capital can be refinanced at lower rates.

Revenue projections are grounded in existing freight volumes currently paying congestion penalties and delay premiums.

The break-even calculation incorporates:

- Direct subscription revenue

- Dynamic capacity revenue

- Energy optimization savings

- Asset appreciation linked to corridor utilization

Importantly, SAN's financial case is strengthened by avoided public expenditure.

Avoided Public Costs

Surface congestion imposes recurring fiscal burdens on governments:

- Highway resurfacing and expansion

- Port yard land acquisition

- Congestion mitigation programs

- Emissions compliance subsidies

- Health costs linked to air pollution

By relocating primary freight flow underground, surface maintenance burdens decline.

Fewer heavy trucks on highways reduce pavement degradation. Port expansion pressure eases. Urban land acquisition slows.

When calculating break-even from a public-sector perspective, avoided capital expenditure and maintenance costs must be included.

The comparison is not between building SAN and spending nothing.

It is between:

- Continuing to finance expanding surface mitigation indefinitely

- Investing once in structural recalibration

When avoided costs are integrated into financial modeling, break-even timelines compress.

Risk Mitigation and Credit Strength
Institutional investors prioritize risk containment.

SAN reduces key risk categories through:

- Long-term subscription contracts

- Diversified corridor user base

- Energy cost stabilization through renewables

- Automation-driven operational consistency

Multi-functional infrastructure value (transport + energy + resilience) Because freight circulation is essential to national economies, demand risk is lower than discretionary infrastructure categories.

Properly structured, SAN corridors can achieve infrastructure-grade credit ratings.

The objective is not speculative growth multiples. It is stable, bond-like performance with strategic national benefit.

Chapter 10

Legislative & Regulatory Pathways

Engineering feasibility and financial logic alone do not deliver infrastructure.
Regulatory clarity does.

Large-scale systems require predictable governance frameworks. Investors require legal certainty. Operators require standardized compliance pathways. Communities require defined protections.

Subterranean freight infrastructure introduces governance questions distinct from surface projects. These questions are not barriers—but they must be addressed directly and systematically.

Clear legislative architecture transforms underground corridors from conceptual possibility into deployable national assets.

Regulatory Certainty as Capital Catalyst

Infrastructure capital flows toward jurisdictions that offer:

- Transparent permitting processes
- Defined property rights
- Stable long-term policy commitments
- Predictable environmental review frameworks

Without regulatory clarity, even financially viable projects face delay risk, litigation exposure, and capital cost escalation.

For SAN to scale, governments must establish a legal environment that recognizes depth as strategic infrastructure territory—just as airspace, highways, and maritime zones are codified today.

The objective is not deregulation. It is structured modernization.

Sub-Surface Easements

Most national legal systems clearly define surface property rights. Subterranean ownership, however, varies significantly across jurisdictions.

Historically, property ownership was often interpreted to extend indefinitely downward. Modern legal systems have already modified this principle to accommodate:

- Utility tunnels
- Metro rail systems
- Sewer networks
- Mining operations
- Energy transmission corridors

These precedents demonstrate that sub-surface governance frameworks can be structured without undermining surface ownership.

Legislative reform for SAN can formalize **depth-based sub-surface easements** that:
- Permit corridor development below specified depth thresholds
- Protect existing surface ownership rights
- Standardize compensation mechanisms where required
- Define access protocols for maintenance and safety

By codifying a clear vertical separation between surface property and deep freight corridors, governments reduce litigation risk and streamline permitting.

Depth thresholds can be established to ensure no interference with building foundations or utilities. Compensation frameworks can be standardized to avoid protracted negotiation processes.

The principle is straightforward: surface rights remain intact; designated depth zones are recognized as public utility corridors.

Such frameworks already exist in various forms. SAN extends them to freight mobility at national scale.

Streamlined Permitting Frameworks

Traditional surface infrastructure projects often encounter complex approval pathways involving:

- Environmental impact assessments
- Community hearings
- Zoning variances
- Multi-agency coordination

Subterranean systems, while technically complex, may reduce certain categories of surface disruption. With fewer relocations, less visible land alteration, and reduced noise exposure, the public impact profile differs significantly from highway expansion projects.

Governments can establish dedicated subterranean infrastructure review processes that:

- Consolidate agency oversight
- Standardize environmental assessment criteria for underground systems
- Establish expedited review timelines for qualifying corridors

Predictable timelines lower financing costs by reducing pre-construction uncertainty.

The regulatory objective is procedural clarity—not acceleration at the expense of safety or environmental protection.

Safety and Operational Standards

Regulatory legitimacy depends on safety transparency.

National standards bodies can codify:

- Tunnel structural integrity requirements
- Fire suppression systems
- Ventilation and air quality thresholds
- Emergency access protocols
- Digital monitoring and cybersecurity safeguards

Existing standards from metro rail systems, highway tunnels, and energy pipelines provide reference models.

SAN does not introduce unknown engineering categories. It integrates established tunneling, electrification, and automation technologies within a freight framework.

Codified safety standards increase public trust and reduce investor risk perception.

Zoning Reform and the Surface Dividend

As freight infrastructure transitions underground, surface zoning policy becomes a strategic lever.

Surface zoning has historically segregated industrial, residential, and commercial uses. Large warehouse districts and container yards were justified by proximity to surface freight routes.

When primary trunk movement relocates underground, zoning can be rebalanced.

Former warehouse districts may transition to:

- Mixed-use urban development
- Transit-oriented housing
- Public green space
- Renewable energy installations
- Flood mitigation infrastructure

Port-adjacent land previously required for overflow storage can be reclassified for higher civic or economic value uses.

Zoning reform does not occur automatically. It requires coordinated planning between municipal governments, port authorities, and regional development agencies.
The surface dividend becomes policy-enabled value rather than incidental surplus.

Interagency Coordination
Freight systems intersect multiple regulatory domains:

- Transportation ministries
- Energy regulators
- Environmental agencies
- Water authorities
- Urban planning departments

Subterranean corridors, particularly when integrated with renewable energy and water reserves, span these domains simultaneously.

Governments may establish dedicated interagency task forces or national corridor authorities to coordinate:

- Permitting
- Environmental compliance
- Energy integration
- Emergency response planning
- Public communication

Centralized coordination reduces duplication and aligns objectives across departments.

Infrastructure integration requires governance integration.

National Adoption Models

National deployment need not begin at full scale.

Adoption can proceed through flagship corridors connecting:

- Major maritime ports
- Primary inland logistics hubs
- High-density manufacturing zones

Early corridors should be selected based on:

- Demonstrable congestion cost
- Strong subscription demand
- Renewable energy alignment
- Clear right-of-way feasibility

These pilot segments establish regulatory precedent and operational proof.

Once performance metrics demonstrate reliability, financial viability, and safety compliance, replication becomes administratively simpler. Legislative frameworks designed for the first corridor can be adapted for subsequent expansions.

Success scales through iteration.

Regional and Cross-Border Alignment

Freight networks cross political boundaries. Regulatory alignment must therefore operate at multiple scales.

Regional trade blocs can coordinate:

- Corridor interoperability standards

- Digital tracking protocols

- Safety and inspection harmonization

- Cross-border customs integration

International standards bodies may establish technical guidelines for:

- Subterranean freight container interfaces

- Automated routing communication systems

- Energy integration protocols

The historical transition to containerization required global standardization of box dimensions and port handling equipment. Subterranean freight corridors require similar coordination—though focused on interoperability rather than physical container design.

Alignment does not require simultaneous global adoption. It begins with regional leadership and expands through demonstration.

Legal Risk Mitigation

Legislative clarity reduces three primary risks:

1. Property litigation risk
2. Permitting delay risk
3. Operational compliance ambiguity

When frameworks are codified in advance, investors can model risk with greater precision. Capital cost declines when regulatory uncertainty declines.

Clarity is not merely administrative efficiency. It is financial leverage.

Public Communication and Political Mandate

Infrastructure projects of this scale require public understanding.

Subterranean freight corridors differ from surface megaprojects in one important respect: their benefits are often less visible.

Reduced truck traffic, improved air quality, reclaimed land, and stabilized energy costs must be communicated clearly.

Governments can frame legislative reform around measurable objectives:

- Congestion reduction targets
- Emission reduction benchmarks
- Land reclamation commitments
- Renewable integration milestones

Policy transparency builds durable political support.

Phased Legislative Strategy

Governments need not overhaul entire legal codes simultaneously.

A phased legislative pathway may include:

Phase 1: Corridor Authorization Act

- Establishes legal recognition of sub-surface freight corridors
- Defines depth-based easement standards
- Authorizes pilot corridor development

Phase 2: Regulatory Harmonization

- Standardizes safety, energy, and environmental frameworks
- Coordinates interagency oversight

Phase 3: Expansion and Replication Framework

- Enables streamlined approval for additional corridors
- Codifies subscription-based infrastructure revenue recognition

Incremental legislative sequencing reduces political resistance and allows regulatory refinement based on operational experience.

PART IV

Strategic Summary

Surface congestion is not an operational inconvenience to be managed. It is a structural penalty embedded within the current geometry of trade.

The Clutter Tax extracts value quietly but consistently. It reduces competitiveness, inflates environmental liabilities, consumes scarce urban land, and constrains GDP potential. Incremental mitigation—widening highways, expanding yards, extending operating hours—treats symptoms. It does not remove the source of friction.

The strategic choice before governments and investors is therefore clear: continue financing horizontal expansion with diminishing returns, or recalibrate the geometry of circulation itself.

SAN reframes that decision in financial terms.

Its viability rests on monetizing reliability, aligning with green capital markets, leveraging public–private partnership frameworks, and deploying capital in disciplined phases. The Subscription-to-Speed model converts congestion relief into recurring, contract-based revenue. Green bonds reduce cost of capital by aligning infrastructure with decarbonization mandates. PPP structures distribute construction and demand risk across public and private stakeholders. A phased 10-year break-even pathway provides clarity to institutional investors.

Break-even is not measured solely through direct user revenue. It includes avoided surface expansion costs, reduced environmental exposure, stabilized energy pricing, and the restored value of reclaimed land. The financial architecture mirrors the engineering architecture: structured, layered, and integrated.

Yet financial feasibility depends on regulatory clarity. And regulatory clarity depends on deliberate legislative modernization.

Subterranean infrastructure does not require unprecedented legal invention. It requires structured adaptation of existing precedents in utilities, tunneling, mining rights, and public–private governance. By formalizing depth-based sub-surface easements, modernizing zoning policy to unlock the surface dividend, codifying safety standards, and aligning national and regional regulatory frameworks, governments transform SAN from engineering blueprint into authorized public utility.

The pathway to adoption is procedural rather than revolutionary:

Begin with flagshipcorridors.
Codify depth-based rights.
Demonstrate safety and financial performance.
Align standards regionally.
Replicate systematically.

In this sequence, structural redesign becomes investable reality.

The next stage of analysis moves beyond corridor economics and regulatory foundations to geopolitical positioning—examining how early-adopting nations can secure durable competitive advantage by redefining the physical geometry of trade before congestion and climate pressures force reactive reform.

Part V

Global Implementation & Long-Term Legacy

Design proves feasibility. Implementation determines legacy.

The Subterranean Arterial Network (SAN) achieves structural impact only when it moves beyond national pilots and becomes coordinated global infrastructure. Engineering viability and financial logic establish the foundation. Standardization, demonstration, and replication establish permanence.

This section shifts from blueprint to scale.

First, it outlines the governance framework required to ensure interoperability across borders. Global trade depends on shared standards; subterranean corridors must operate within the same disciplined alignment. Second, it examines a prototype corridor as measurable proof—where congestion relief, emission reduction, and surface recovery are quantified rather than projected. Finally, it considers the long-term civic and economic implications when freight circulation no longer consumes the surface.

Infrastructure defines spatial order for generations.

If depth becomes the new geometry of trade, the legacy will not be limited to faster movement. It will be measured in reclaimed land, stabilized climate exposure, interoperable global corridors, and cities that regain flexibility above strengthened arteries below.

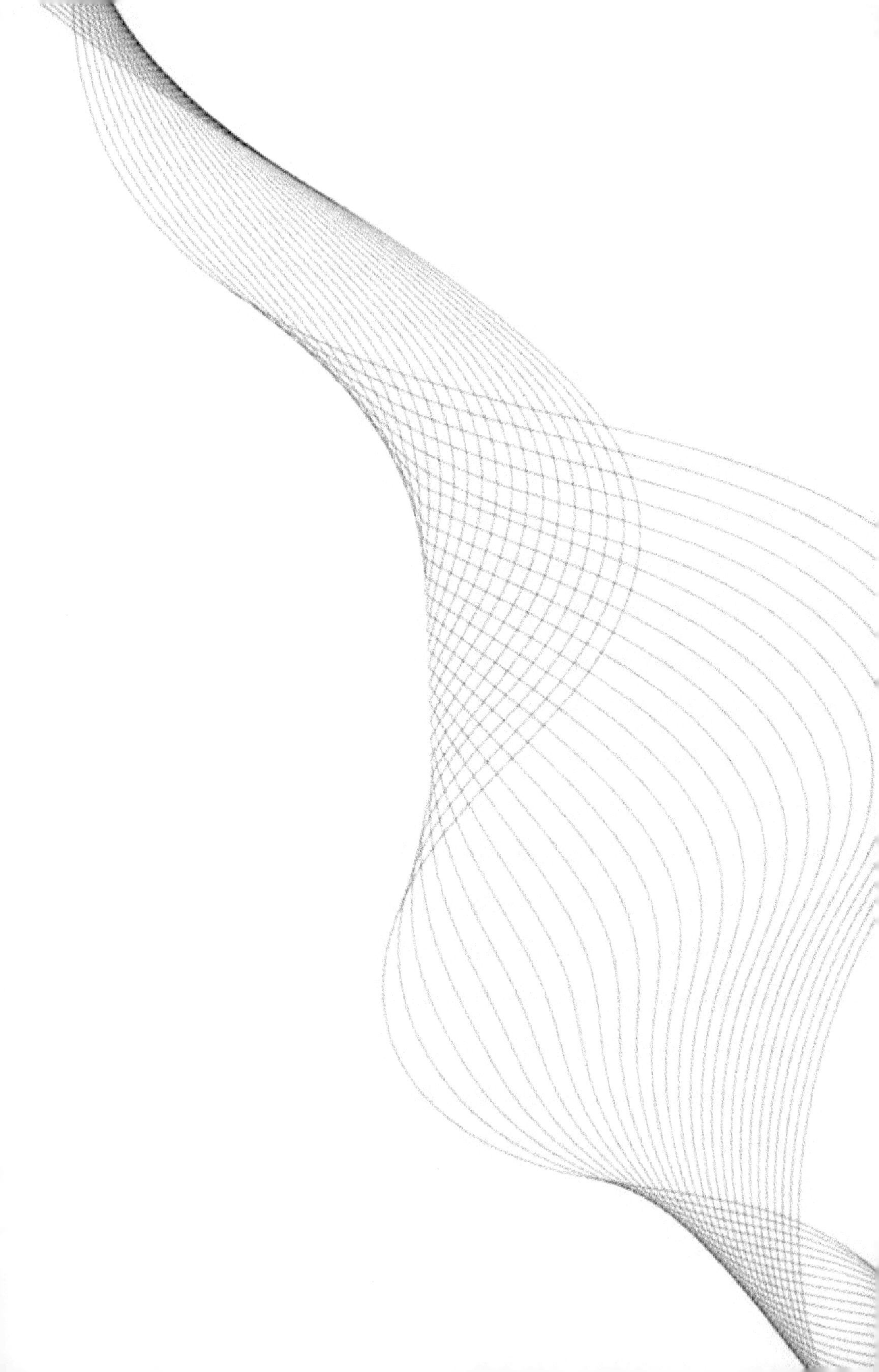

Chapter 11

The Jagir Protocol

Global trade does not function on ambition alone.

It functions on agreement.

Containerization transformed commerce because nations adopted common dimensions, handling systems, and maritime rules. A container loaded in one country could be unloaded in another without redesign, negotiation, or mechanical adjustment. Interoperability—not invention alone—created scale.

The Subterranean Arterial Network (SAN) requires the same discipline.

Engineering corridors within a single nation is achievable. Scaling them across borders demands structured alignment. Without shared standards, fragmented systems would emerge—raising cost, increasing risk, and undermining investor confidence.

The Jagir Protocol establishes the governance architecture and technical standards necessary to scale SAN internationally while preserving compatibility, investment security, and operational reliability.

It is not a technological breakthrough.

It is a coordination framework

The Logic of Standardization

Infrastructure that crosses borders must operate within predictable parameters. Freight operators, shipping lines, insurers, and infrastructure funds all depend on system continuity.

Standardization delivers:

- Interoperability across jurisdictions
- Reduced capital risk
- Lower insurance premiums
- Predictable equipment manufacturing
- Accelerated deployment timelines

Without shared standards, each corridor becomes a bespoke project. Customization inflates cost and limits network effect.

With shared standards, scale compounds.

The Jagir Protocol codifies those shared parameters.

International Technical Specifications

To enable global replication, SAN corridors must adhere to standardized specifications across core engineering domains.

1. Tunnel Geometry Standards
- Defined diameter ranges
- Structural reinforcement thresholds
- Pressure-resistant lining specifications
- Drainage and flood mitigation criteria

Standardized geometry ensures that rolling platforms, maglev systems, and container modules operate seamlessly across national segments.

2. Pressure and Reduced-Environment Parameters
- Maximum and minimum pressure ranges
- Emergency equalization protocols

- Airlock sequencing standards
- Redundancy systems for pressure stability

Reduced-pressure operation enables high-speed movement with lower aerodynamic drag. Consistent parameters ensure container modules do not require reconfiguration when crossing borders.

3. Maglev Interface Standards

- Track width and magnetic alignment specifications
- Power supply voltage ranges
- Control signal compatibility
- Fail-safe braking integration

Magnetic levitation propulsion systems must adhere to harmonized interface dimensions to prevent mechanical incompatibility between national networks.

4. Container Carriage Modules

- Load-bearing platform dimensions
- Locking mechanisms compatible with ISO container standards
- Vibration tolerances
- Automated identification tagging systems

The objective is simple: a container entering the network in one country must travel through corridors in another without redesign or mechanical adaptation.

5. Digital Routing Protocols

- Encrypted communication standards
- Real-time logistics data integration
- AI-driven traffic allocation compatibility
- Cybersecurity requirements

Digital interoperability is as critical as mechanical compatibility. Routing systems must communicate seamlessly across jurisdictional boundaries to preserve throughput stability.

6. Safety and Redundancy Benchmarks

- Emergency evacuation protocols
- Fire suppression integration
- Structural stress monitoring
- Independent power backup systems

Standardized safety frameworks reduce regulatory duplication and accelerate insurance underwriting.

International transport bodies, trade blocs, and standards organizations would collaborate to codify these specifications. The model mirrors established global frameworks in maritime law, aviation safety, and telecommunications standards.

Global utility depends on alignment.

Standardization reduces risk.

It lowers capital cost.

It accelerates adoption.

The Universal Docking Interface (UDI)

If the corridor is the artery, the interface is the valve.

The Universal Docking Interface (UDI) becomes the standardized physical bridge between surface ports and subterranean corridors.

The UDI defines:

- Vertical transfer shaft dimensions
- Container descent and ascent mechanisms
- Automated locking systems
- Identification and routing confirmation protocols
- Energy and data connectivity interfaces

By defining a universal interface, ports can integrate SAN compatibility incrementally.

A port does not need to redesign its entire footprint in a single capital cycle. Instead, designated berths can be retrofitted with compliant docking shafts over time. As throughput shifts underground, additional berths can be converted.

The UDI protects long-term investment by ensuring:

- Equipment purchased today remains compatible with future corridor expansions
- Ports avoid stranded assets
- Infrastructure funds retain asset liquidity

Just as standardized container cranes transformed global shipping, the UDI standardizes the transition from surface congestion to vertical circulation.

Governance Structure of the Jagir Protocol

The Jagir Protocol operates at three governance levels:

National Level
- Adoption of standardized engineering codes
- Formal recognition of sub-surface freight corridors
- Regulatory alignment with safety and digital protocols

Regional Level
- Harmonized cross-border interoperability standards
- Customs data integration
- Coordinated infrastructure financing initiatives

International Level
- Technical specification codification
- Safety benchmarking
- Dispute resolution frameworks
- Cybersecurity coordination

This layered governance ensures that no single nation controls the network, yet all participants adhere to common operating principles.

Investment Security Through Alignment

Capital markets reward predictability.

When infrastructure conforms to internationally recognized standards:

- Credit ratings improve
- Insurance costs decline
- Equipment manufacturing scales efficiently
- Private capital participation expands

Fragmentation increases financing cost. Alignment reduces it.

The Jagir Protocol transforms SAN from a national infrastructure experiment into a standardized global asset class.

Global Adoption Strategy

Implementation follows a disciplined, phased model.

Phase 1 – Strategic National Corridors

High-density port-to-inland routes where congestion costs are highest. These corridors demonstrate:

- Economic viability
- Operational reliability
- Regulatory clarity
- Subscription-based revenue stability

Performance metrics from these corridors provide empirical validation.

Phase 2 – Regional Integration

Corridors extend across neighboring economic zones. Trade blocs coordinate standards, enabling cross-border freight acceleration.

Interoperability across two or three adjacent nations establishes proof of multinational compatibility.

Phase 3 – Global Network Alignment

Major trade corridors connect across continents. Subterranean arteries beneath primary economic regions begin to interlink.

The network effect strengthens:

- Reliability improves
- Insurance risk declines
- Investment capital expands
- Competitive pressure encourages late adopters to participate

Adoption does not require universal participation at the outset.

It scales through demonstrated return,regulatory clarity, and competitive advantage.

Competitive Dynamics

Infrastructure leadership generates economic leverage.

Nations that adopt early gain:
- Throughput stability
- Reduced congestion volatility
- Lower logistics costs
- Reclaimed surface land
- Renewable energy integration advantages

Manufacturers gravitate toward predictable supply chains. Investors favor resilient infrastructure environments. Insurance markets reward reduced risk exposure.

Market forces encourage replication.

The Jagir Protocol ensures that replication occurs within a coherent framework rather than fragmented experimentation.

Chapter 12

Prototype Model – Sacramento Corridor

Large-scale transformation requires visible demonstration.

Policy frameworks, financing models, and international protocols establish feasibility. Yet infrastructure of this magnitude must ultimately be validated through operation. A working corridor generates data, public confidence, and investor clarity.

The Sacramento Corridor model illustrates how the Subterranean Arterial Network (SAN) can operate within a defined geography—linking a major coastal port to an inland logistics zone through a dedicated subterranean artery.

It is not a symbolic pilot.

It is a measurable prototype.

Geographic Logic of the Corridor

The Sacramento Corridor would connect a high-volume maritime terminal on the California coast to an inland distribution hub near Sacramento. The alignment would follow, where feasible, existing transportation rights-of-way beneath established highway corridors.

This route presents strategic advantages:

- Significant container throughput from Pacific trade flows
- Chronic surface congestion along major freight highways
- Established inland warehouse and distribution clusters
- Renewable energy generation potential in adjacent regions
- A state-level policy environment supportive of decarbonization

The corridor would function as a direct trunk artery beneath the surface grid. Containers discharged at port would descend through Deep-Well Docking shafts into the underground network. From there, they would travel at high speed to inland Sub-Hubs.

Transit that currently requires extended trucking cycles would occur within predictable, scheduled windows.

The objective is not incremental improvement.

It is structural bypass.

Port-to-Inland Demonstration
The operational sequence is straightforward:

1. A container vessel docks at a compliant berth.
2. Containers are transferred to vertical descent platforms via the Universal Docking Interface.
3. Containers enter the reduced-pressure subterranean corridor.
4. Within hours, freight arrives at the Sacramento Sub-Hub.
5. Automated sorting systems allocate cargo to regional distribution channels.

Surface trucking is reserved for regional and last-mile distribution rather than long-haul port clearance.

This demonstration focuses on measurable outcomes:

- Reduced truck volume on primary highways
- Shortened container dwell time at port
- Stabilized delivery windows for manufacturers and retailers
- Lower per-unit emission intensity
- Improved labor scheduling predictability

A prototype corridor is not a marketing exercise. It is data-generating infrastructure.

Traffic Reduction Modeling

A central metric for the Sacramento Corridor is truck displacement. Baseline modeling would establish:

- Average daily container volume leaving port by truck

- Peak-hour freight density on primary highway segments

- Historical congestion patterns and delay frequency

Diversion scenarios would then project phased migration of container flows into the subterranean corridor. Even partial diversion—such as 30–50 percent of long-haul freight—could materially reduce peak-hour truck density.

The implications extend beyond freight:

- Lower roadway wear from heavy axle loads

- Reduced accident risk

- Improved commuter travel time reliability

- Decreased emergency response congestion

Public agencies would quantify deferred highway expansion projects and extended pavement lifespan.
Traffic modeling must incorporate:

- Baseline freight volume

- Projected subscription adoption rates

- Diversion percentages by cargo category

- Emission reduction estimates

- Road maintenance cost savings

- Insurance cost implications linked to reduced accident frequency

Quantification builds political and investor confidence. Measured results replace theoretical promise.

Emission and Energy Performance

The Sacramento Corridor would also provide clear decarbonization metrics.

Baseline analysis would assess:

- Diesel consumption from displaced trucking

- Idle emissions at port staging areas

- Emission intensity per container-mile

Post-implementation measurement would compare:

- Electrified corridor energy consumption

- Renewable energy integration percentage

- Lifecycle emission intensity per container

Reduced-pressure transport and maglev propulsion—powered through renewable-aligned grids—would materially lower per-unit emissions.

The result is not dependent on behavioral mandates. It arises from structural reallocation of freight movement.

The prototype would therefore produce real-world emission data suitable for:

- Carbon accounting validation

- Green bond certification

- ESG investment reporting

- Climate policy benchmarking

Inland Sub-Hub Integration

Near Sacramento, the inland Sub-Hub would serve as the redistribution heart of the corridor.

Functions would include:

- Robotic container sorting

- Automated routing allocation

- Modular breakdown for regional distribution

- Renewable-powered storage systems

- Integrated water reserve chambers where applicable

The Sub-Hub reduces the need for expansive surface warehouse sprawl by concentrating throughput below grade and limiting surface staging to final distribution cycles.

Operational metrics would track:

- Container throughput per hour

- Sorting cycle time

- Energy consumption per unit

- Labor efficiency improvements

- Maintenance cycle frequency

Performance transparency strengthens investor and regulatory confidence.

Urban Surface Transformation

One of the most visible outcomes of the Sacramento Corridor would be surface land recovery.

At the coastal port, overflow container yards and truck staging areas could gradually be reduced as vertical extraction replaces horizontal accumulation.

In inland zones, expansive surface storage acreage could be consolidated as underground flow stabilizes throughput.

Urban planners, in coordination with municipal authorities, could reallocate reclaimed land toward:

- Affordable housing development

- Transit-oriented infrastructure

- Public green space

- Renewable energy installations

- Flood mitigation systems

The Sacramento Corridor would demonstrate the surface dividend in tangible form.

Land once consumed by congestion becomes civic capital.

Economic Multiplier Effects

A functioning corridor would influence broader economic behavior. Manufacturers would gain:

- Predictable inland delivery windows

- Reduced inventory buffer requirements

- Lower insurance premiums linked to transit reliability

Retailers would benefit from:

- Stable supply chain scheduling

- Reduced stockout risk

- Improved pricing predictability

State-level agencies would capture:

- Deferred infrastructure maintenance expenditure

- Improved air quality metrics

- Increased taxable land value from redevelopment

The multiplier effect extends beyond transport efficiency. It influences investment confidence and regional competitiveness.

Phased Implementation Timeline
The Sacramento Corridor would follow a structured development sequence:

Phase 1 – Feasibility & Environmental Review

- Engineering surveys

- Depth analysis

- Environmental impact modeling

- Regulatory alignment

Phase 2 – Construction of Core Segment

- Port docking shafts

- Initial corridor bore

- Inland Sub-Hub installation

Phase 3 – Operational Launch

- Anchor tenant subscription agreements

- Gradual freight diversion

- Performance monitoring

Phase 4 – Expansion & Optimization

- Capacity scaling

- Additional docking interfaces

- Integration with renewable energy clusters

This phased model reduces risk and allows real-time calibration.

Risk and Resilience Testing

The prototype also functions as a resilience stress test.

Operational modeling would simulate:

- Surface wildfire disruptions

- Flood events

- Highway closures

- Energy grid fluctuations

By maintaining freight continuity underground, the corridor would demonstrate practical resilience benefits beyond congestion reduction. Resilience metrics become part of the financial case.

Strategic Significance

The Sacramento Corridor is not an isolated regional improvement. It is a scalable template.

If validated, its operational data would inform:

- National corridor expansion

- Federal infrastructure planning

- International standardization under the Jagir Protocol

- Green capital allocation strategies

Success does not depend on universal adoption at inception.

It depends on proof.

Demonstrated throughput stability, emission reduction, traffic relief, and land reclamation create a replicable case model.

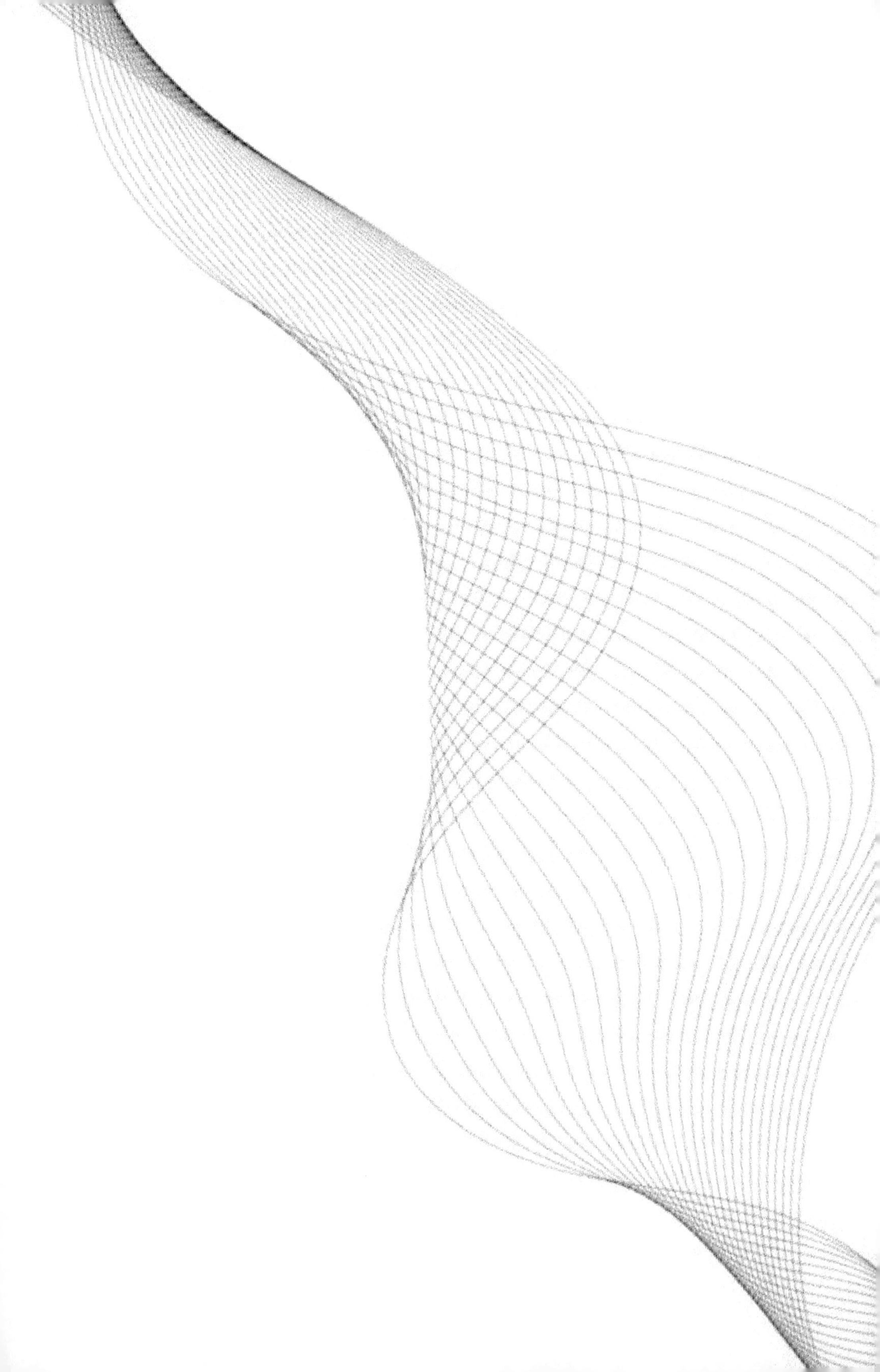

Chapter 13

The Surface Reclaimed

Infrastructure decisions shape cities for generations.

Highways drawn through neighborhoods in the mid-20th century still define urban form today. Rail corridors influence industrial clustering decades after construction. Ports determine air quality patterns for entire metropolitan regions.

The Subterranean Arterial Network (SAN) is not solely a freight velocity system. Its long-term legacy extends beyond container throughput and logistics efficiency.

It reshapes who benefits from land, air quality, economic access, and urban flexibility.

If earlier chapters focused on arteries, speed, and structural recalibration, this chapter addresses the civic dividend.

When freight descends, the surface reopens.

Environmental Justice Through Structural Redesign

Port-adjacent communities across the world share a common pattern:

- Elevated diesel emissions
- Concentrated truck traffic
- Persistent industrial noise
- Higher rates of respiratory illness
- Limited access to green space

These communities frequently did not choose their proximity to heavy logistics infrastructure. They inherited it through historical zoning decisions, industrial clustering, and economic necessity.

Traditional mitigation strategies—grant programs, emission retrofits, or buffer landscaping—treat symptoms. They reduce harm incrementally but leave the primary source of pollution intact: surface freight concentration.

Environmental justice is not achieved through isolated mitigation. It requires structural redesign.

By relocating primary freight arteries underground and electrifying propulsion systems, SAN addresses the source.

Reduced surface trucking lowers particulate concentration in residential corridors. Electrified movement eliminates localized diesel combustion within primary trunk routes. Concentrated staging yards shrink as vertical extraction replaces horizontal accumulation.

The impact is measurable:

- Fewer heavy-duty trucks idling near homes and schools

- Reduced nitrogen oxide and particulate emissions

- Lower ambient noise levels

- Improved predictability in neighborhood traffic flow

Environmental benefit shifts from theoretical to structural.

When freight bypasses surface streets, pollution burdens redistribute more equitably across metropolitan regions.

This is not redistribution by mandate. It is redistribution by design.

Air Quality and Public Health

Air quality improvements carry direct economic implications.

Public health expenditures tied to asthma, cardiovascular disease, and pollution-related illness represent recurring fiscal costs for governments. Reduced heavy diesel concentration near dense populations lowers long-term healthcare strain.

Electrified corridors, integrated with renewable energy sources, further reduce lifecycle emissions per container-mile. As adoption scales, metropolitan air baselines improve incrementally but persistently.

Health outcomes do not improve overnight. They improve through consistent exposure reduction.

Infrastructure is one of the few policy levers capable of delivering that consistency at scale.

Urban Redesign and Spatial Liberation

Surface reclamation unlocks planning flexibility that has been constrained for decades.

Cities historically shaped around freight corridors often orient development away from industrial zones. Container yards, truck depots, and overflow lots create barriers—physical and psychological—between neighborhoods and waterfronts.

When staging areas migrate underground, the spatial equation changes.

Reclaimed land can support:

- Mixed-use housing developments

- Transit-oriented districts

- Public parks and waterfront access

- Renewable energy installations

- Flood mitigation basins

- Community health and recreation facilities

Urban redesign becomes a direct dividend of subterranean transition.

The surface dividend is not abstract. It appears in land use maps, zoning reclassification, and redevelopment plans.

Land once reserved for logistical overflow becomes available for human-centered activity.

Economic Access and Equity

Freight infrastructure influences economic opportunity distribution.

When logistics sprawl consumes valuable urban acreage, it constrains housing supply and inflates land prices. When truck corridors dominate certain districts, small businesses and pedestrian commerce decline.

By consolidating trunk movement underground, SAN enables more balanced economic development.

Reclaimed land near ports and inland hubs may support:

- Workforce housing

- Innovation districts

- Public transportation hubs

- Local business corridors

Increased housing supply near employment centers reduces commuter travel distances and strengthens regional labor markets.
The economic access effect is indirect but powerful: spatial flexibility expands opportunity.

Climate-Resilient Urban Fabric

Subterranean transition also supports climate adaptation.

Freed surface land can be reallocated toward:

- Green corridors that absorb heat

- Tree canopy expansion to mitigate urban heat islands

- Water retention systems to manage storm surge

- Elevated transit systems integrated with resilient design

Cities gain the ability to redesign surface space not around freight pressure, but around climate resilience.

The result is a layered urban fabric:

- Circulation below

- Community above

- Resilience integrated throughout

The arterial metaphor completes its cycle: when circulation improves, the body regains vitality.

Decoupling Growth from Land Expansion

Over a 100-year horizon, the most significant implication of SAN is structural decoupling.

Historically, trade growth has required parallel expansion of surface infrastructure—wider highways, larger yards, more warehouses.

This model is land-intensive and environmentally escalatory.

SAN enables a different trajectory.

Trade growth can continue while surface land consumption stabilizes or declines.

Freight electrification accelerates national decarbonization pathways. Urban cores regain land previously dedicated to logistics storage. Energy and transport systems integrate more seamlessly. Climate resilience strengthens through hardened subterranean networks.

Growth no longer requires horizontal sprawl.

A 100-Year Global Outlook

Infrastructure built today will shape economic patterns for the next century.

Over a 100-year horizon, several long-term outcomes become plausible:

1. **Trade scalability without proportional land expansion** Subterranean arteries absorb increased volume without expanding surface congestion.
2. **Integrated energy-transport ecosystems** Electrified corridors align freight movement with renewable generation cycles.
3. **Reduced cumulative emission burden** Decades of electrified throughput compound climate benefit.
4. **Urban land recovery across major port cities** Waterfronts and logistics districts transition toward civic and economic diversification.
5. **Strengthened resilience during environmental volatility** Hardened corridors maintain economic continuity despite climate stress.

The legacy of SAN is not defined by speed alone.

It is defined by spatial liberation.

Intergenerational Value

Infrastructure decisions today will outlast current political cycles and investment horizons.

Highways constructed in the 1950s still shape cities. Ports expanded in the 1970s still define waterfront land use.

The question for this generation is whether trade modernization continues to consume surface space—or whether it restores it.

SAN represents a choice to:

- Preserve economic dynamism
- Reduce environmental escalation
- Recover urban flexibility
- Align freight mobility with long-term climate reality

Intergenerational value emerges when infrastructure solves multiple constraints simultaneously.

PART V

Strategic Summary

Global trade has always depended on standardized arteries.

In the 20th century, the shipping container aligned ports, vessels, and rail systems under a common geometry. In the 21st century, subterranean freight corridors require an equivalent institutional spine.

The Jagir Protocol provides that spine.

It establishes the governance and technical framework necessary to scale underground freight infrastructure internationally while protecting interoperability, capital security, and operational continuity. Its function is straightforward but foundational:

- Standardized tunnel geometry
- Harmonized maglev propulsion interfaces
- Universal docking integration
- Aligned digital routing and data protocols
- Codified safety and redundancy benchmarks

The innovation is not solely in engineering depth. It is in institutional coordination.

When standards align, adoption accelerates
When adoption accelerates, scale compounds.
When scale compounds, congestion's structural penalty declines.

The Sacramento Corridor translates that governance architecture into measurable reality.

It embodies the transition from conceptual design to operating infrastructure. Within a defined geography, it demonstrates:

- Structural congestion relief through truck displacement

- Electrified high-speed trunk movement beneath existing corridors

- Quantifiable emission reduction per container-mile

- Surface land recovery at port and inland nodes

- Financial viability through subscription-based throughput models

A prototype corridor is not symbolic. It is empirical.

When freight descends immediately rather than dispersing horizontally, circulation stabilizes. When arteries operate beneath the surface grid, pressure dissipates above it.

The Sacramento Corridor therefore becomes a working illustration of the surface dividend—where economic circulation strengthens below ground and civic value expands above it.

That civic value is neither abstract nor rhetorical. It is the cumulative outcome of structural redesign.

When primary freight arteries relocate underground:

- Environmental burdens decline at the source

- Urban land becomes available for housing, transit, and public space

- Economic opportunity distribution broadens

- Climate resilience strengthens through hardened networks

- Trade growth decouples from horizontal land expansion

The surface dividend becomes visible in air quality metrics, redevelopment plans, infrastructure budgets, and long-term fiscal stability.

Infrastructure determines who benefits from land and who bears the cost of circulation. For decades, surface freight expansion has concentrated burden while diffusing economic gain.

By redesigning the geometry of trade—standardized through the Jagir Protocol, validated by the Sacramento Corridor, and scaled through interoperable corridors—SAN redistributes those benefits more equitably while preserving economic vitality.

This is not a speculative reimagining of trade. It is a coordinated recalibration of its physical architecture.

Global standards provide the framework
Prototype corridors provide the proof.
Surface reclamation provides the dividend.

Together, they form a long-term blueprint in which trade remains scalable, cities regain flexibility, and circulation strengthens without escalating environmental cost.

Conclusion

The Arterial Shift: Reclaiming the Surface by Moving Global Trade Underground
Mahendra Jagir

Global trade has always relied on arteries. For centuries, those arteries have been laid across the surface—roads widened, rails extended, ports expanded, warehouses multiplied. Each expansion increased capacity, but also increased friction: congestion, emissions, land consumption, and climate exposure.

The central thesis of The Arterial Shift is straightforward: reclaiming the surface requires relocating primary freight circulation beneath it. When the main arteries of trade descend underground, the structural pressure imposed on cities, ports, and highways declines. Circulation strengthens below ground. Civic value expands above it.

The Subterranean Arterial Network (SAN) is not a speculative technology. It is an architectural integration of systems that already exist—advanced tunneling, electrified propulsion, renewable energy generation, grid-scale storage, digital logistics platforms, and climate-resilient engineering. The advancement lies in alignment. Geometry replaces expansion. Coordination replaces sprawl.

The environmental dividend is structural rather than symbolic. Electrified corridors reduce lifecycle freight emissions. Renewable integration stabilizes energy sourcing. Surface truck volumes decline. Idle vessel time compresses. Land previously consumed by container yards and warehouse overflow becomes available for housing, public space, renewable installations, and ecological restoration. Emission reduction is achieved not through restriction, but through redesign.

The economic dividend is equally measurable. Predictable energy demand enables long-term renewable contracts and cost stability. Congestion relief reduces delay penalties embedded within supply chains. Subscription-based freight corridors introduce recurring revenue models compatible with institutional capital. Surface land recovery expands municipal tax bases and long-term fiscal capacity. Trade growth decouples from horizontal land consumption.

The societal dividend follows from infrastructure logic. When freight arteries no longer dominate surface corridors, cities regain flexibility. Climate resilience strengthens as core logistics systems operate within controlled environments insulated from wildfire, flood, and heat volatility. Water reserves, fire suppression systems, and disaster logistics can be embedded within the same corridor spine. Infrastructure becomes layered, serving economic circulation and civic stability simultaneously.

This is a generational shift in spatial design.

Just as railroads reorganized continents and highways reshaped metropolitan form, subterranean freight corridors redefine the geometry of trade for a climate-constrained century. The decision is not whether trade will expand—it will. The strategic question is whether expansion continues to consume the surface or transitions into depth-based circulation that preserves it.

SAN positions depth as the new backbone of global commerce.

Implementation will require disciplined governance, interoperable standards, public–private coordination, and phased demonstration corridors. None of these requirements exceed current institutional or technological capacity. The tools are available. The engineering is proven. Capital markets are actively seeking resilient, long-duration infrastructure assets. What remains is strategic alignment.

Infrastructure determines who benefits from land and who bears the cost of movement. By relocating freight below grade, the burden of circulation is reduced and the value of land is redistributed. The surface dividend—cleaner air, reclaimed acreage, stabilized climate exposure, and strengthened fiscal systems—emerges as a byproduct of structural redesign.

Reclaiming the surface is not an aspiration. It is an engineering choice.

When the primary arteries of trade move underground, cities do not lose vitality—they gain room to breathe, build, and adapt. Economic velocity is preserved. Environmental intensity declines. Civic flexibility expands.

The arterial shift is therefore not a departure from modern trade. It is its maturation.

The next era of global infrastructure will be defined not by how far we extend the surface, but by how intelligently we use depth.